ROMAN
SHIELDS

ABOUT THE AUTHORS

Hilary Travis holds a Masters and Honours Degrees in Archaeology. In addition to over twenty years' experience as an archaeologist, she also has over twenty years' combat experience in Japanese martial arts, and over ten years in reconstruction of Roman and medieval period artefacts.

John Travis is an established author. His first book, Coal in Roman Britain, was based on his PhD thesis. He holds a Masters Degree and Doctorate in Roman Archaeology from the University of Liverpool. He is an archaeologist with over thirty years' experience, and an Associate member of the Institute of Field Archaeologists (AIFA).

Both Hilary and John are active re-enactors, as members of both the Chester Guard (Deva VV) Roman Society and the Thomas Stanley Retinue (Wars of the Roses Medieval group).

ROMAN SHIELDS

HISTORICAL DEVELOPMENT AND RECONSTRUCTION

HILARY & JOHN TRAVIS

AMBERLEY

This edition published 2015

Amberley Publishing
The Hill, Stroud
Gloucestershire, GL5 4EP

www.amberley-books.com

British Library Cataloguing in Publication Data.
A catalogue record for this book is available from the British Library.

ISBN 978 1 4456 5523 9 (paperback)
ISBN 978 1 4456 3843 0 (ebook)

Typeset in 10pt on 12pt Sabon.
Typesetting and Origination by Amberley Publishing.
Printed in Great Britain.

CONTENTS

LIST OF TABLES

LIST OF ILLUSTRATIONS

LIST OF COLOUR PLATES

ACKNOWLEDGEMENTS

This book has been the result of a team effort with my husband, J. R. Travis, who has provided most of the photography and illustrations, along with the provision of on-tap supplies of coffee and brain-inspiring nibbles.

I should also like to thank the members of the Deva VV (Chester Guard) Roman Living History Group, who agreed to be photographed in their various first-century AD personas, and some of whom may also recognise themselves in the artist's representations. I should like to acknowledge the assistance of Dr Philip Freeman, of the University of Liverpool, for his encouragement and ceaseless editing and re-editing of the original thesis, one small chapter of which formed the inspiration for this text.

I should also like to express my special thanks to Marcus Vlpivs Nerva (Martin McAree) and other members of Legion Ireland Roman Group, for their invaluable assistance in providing photographs of their reconstructed shields and of their simulated combat use.

For the most part, however, this book was produced as a means to keep me out of mischief and to save me from terminal boredom, during an infinitely forgettable year of serious illness, many weeks of which were spent as a reluctant guest of several local hospitals. I should therefore like to dedicate this book to the doctors and nursing staff at Southport District General Hospital and at Liverpool Heart & Chest Hospital, without whom I would not have been here to bore you all silly with the following work.

I should also like to thank our many friends within the re-enactment community for their well wishes and support during my illness, particularly the Poor Knights of St Dysmas, who made welcome and pampered an invalid at their camp for a few escapist Medieval weekends.

I

INTRODUCTION

BACKGROUND TO THE RESEARCH

The conventional public view of the Roman army comes from a number of sources: art, early filmography (the history according to Hollywood) and romantic literature. The academic viewpoint differs from these popular views, based on more solid, tangible evidence: ancient literature, sculptures, wall paintings and archaeological evidence. In more recent years this has produced more reliable, evidence-based views of the Roman military, as portrayed in museums and by living history re-enactments.

Nevertheless, the conventional view still focussed on the legionary soldier of the Imperial period, following the stereotypical image of uniformity, as seen depicted on Trajan's Column, with all legionaries equipped in segmented iron cuirass (*lorica segmentata*) and auxiliaries in mail (*lorica hamata*) or scale (*lorica squamata*), carrying their standard associated shield type, although with the main focus being the armour assemblage itself, their shield only a secondary concern. The two main shield forms (Fig. 1) that appear to have been used were the semi-cylindrical, rectangular shield, more usually associated with legionary troops (although in the case of the sculpture from Alba Iulia, if one interpretation is correct, possibly also used by gladiators; Fig. 2), and the long oval shield, associated with auxiliary and cavalry units.

However, the reality appears to have been somewhat different, in light of the information on design, construction, developmental progression and use of shields by the Roman military offered from a variety of sources – literary, sculptural and archaeological. These suggest that there was a range of shield shapes in use, many contemporaneously, serving a variety of military functions. In addition, these basic functional types further exhibit developmental changes over time as they adapt to changing fighting styles, enemy weaponry and external cultural influences. Furthermore, the shield's shape and decoration may have served to define to which military unit the bearer belonged, along with his position and function within that unit.

This book, therefore, will take a more in-depth view of the Roman shield, to discuss its origins and developmental progression in shape and function.

While reviewing the standard legionary shield in use, through living history re-enactors in simulated combat situations, factors became evident which would have implications for construction methods, combat use, wear damage, repair and replacement rates. The shields used by modern Roman re-enactors are made from modern materials to produce a shape as close as possible to the perceived original. This is achieved by gluing and bending two sheets of three-layered plywood, either over a former model or in a convex press, screwed together to retain their position while the glue hardens. When dried, the glue helps to prevent the two boards returning to their flat, original shape, a task further aided by a framework of shaping strips applied to the rear. An applied trim of brass strip then protects the shield edge, again helping to prevent separation of the ply layers. The resulting shield then consists of at least six layers of ply, with an overall thickness in excess of 1 cm. While the shape may approximate from a distance the shields seen on Trajan's Column, it would not perform the same in use. The wood and glues used in the modern ply would not exhibit the same properties as those used in the original, and the weight of the reproduction may be considerably heavier,

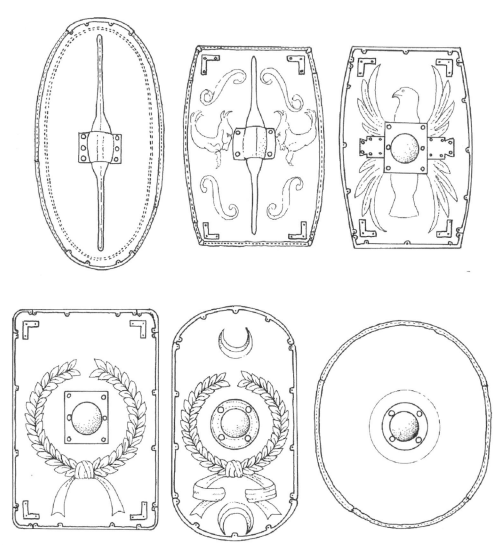

Fig 1. Progression of Roman 'legionary' shield shapes, from Republican period through to late Empire. (Artwork by J. R. Travis)

with implications for how long an army could comfortably march carrying such a shield and how easily it could be wielded in combat situations.

Questions then come to mind as to how the original shield would have been constructed. Clearly the Roman craftsman would not have had access to modern, pre-formed plywood or modern contact adhesive, nor would he have had modern screws to hold the ply in place. He would, however, have had a greater knowledge of the properties of different wood types and access to a greater variety of raw materials, along with many years of acquired, time-served experience of his trade (whereas the average re-enactor is merely an enthusiastic amateur).

This book does not claim to offer a definitive answer to how a real Roman shield was produced, but it does offer a range of possibilities, testing these and discussing their respective feasibilities. The first stage in any such discussion must be to fully research any remaining evidence for construction in antiquity: location of manufacture, materials used and method of construction. The sources of this evidence may come from written descriptions in ancient literature, visual representations in artwork or sculpture, or tangible archaeological evidence. Unfortunately, although many sites have produced evidence of shield furniture (bosses, handles, edging strips, etc.; Figs 3–5), shield boards are generally made from organic materials that do not survive under most conditions. Our physical evidence is therefore limited, although a few notable exceptions do exist which provide some insight into both construction and decoration across many centuries of use. Subsequent chapters will therefore discuss the sources of evidence in an attempt to answer these questions of development, construction, decoration, use and longevity. Some possible theoretical methods of construction will then be tested, using materials as close as possible to those which would have been available, to 'field test' the resulting reconstructions in order to approximate wear damage. By close examination of the published reports detailing the archaeological examples selected, some alternate interpretations are also proposed, with implications for dating, identity of ownership and possible evidence of damage/repair.

Fig 2. Sculpture from Alba Iulia depicting soldier or gladiator wearing composite segmented armour and carrying rectangular legionary-style shield. (Artwork by J. R. Travis, from Stephenson, 2001, fig. 14)

Fig 3. Sword and shield mountings from Newstead. (Photograph from Curle, 1911)

Fig 4. Shield handgrip from the Lunt. (Artwork by J. R. Travis)

LITERARY SOURCES

There appear to be a variety of terms used to describe shields, some referring to specific types and others being used more as terms for shields in general. For example, the term *scutum*, although for the most part used to refer to the semi-cylindrical rectangular form usually associated with legionary troops, may have been used in antiquity as a more general term for any shield type, and by the fourth century AD it had come to refer to broad ovals like those widely used at Dura Europos. The term *hoplon* was also used to describe a shield of unclear typology that may have been similar to its Greek predecessor, which was bossless, round and gripped by passing the hand and forearm through two looping straps on the reverse.

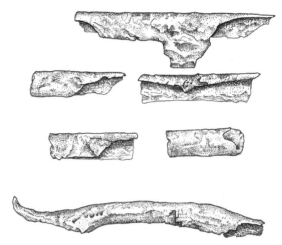

Fig 5. Shield edging strip from the Lunt. (Artwork by J. R. Travis)

Both Livy in the Augustan period and Ammianus in the fourth century AD use the terms *clipeus* and *parma* to refer to shields other than the large, curved, semi-cylindrical, rectangular *scutum*, and may be referring to auxiliary and cavalry shields (Livy, 8.8.3; Ammianus 29.5.39, 31.5.9; Fig. 6). The lightweight, decorated cavalry parade shield is also descibed by Arrian (*Ars tactica* 34.5), writing in the first century AD. Although not himself of Roman origin, Josephus used a variety of terms for shields: *aspis*, being the round/oval shield carried by the *pedites singulares*; *thureos*, being the large, flat oval cavalry shield; and the *thureos epimenes*, being the rectangular shield of the 'rest of the phalanx' (Josephus, *Bell. Iud.* 3.94–7). This latter term may yet again be referring to the semi-cylindrical, rectangular shield more commonly associated with the legionary troops as the *scutum*.

Pliny the Elder described the best timbers to use in the construction of shields as

the most flexible and consequently the most suitable for making shields, are those in which an incision draws together at once and closes up its own wound, and which consequently is the more obstinate in allowing steel to penetrate; this class contains the vine, *agnus castus*, willow, lime, birch, elder and both kinds of poplar. Of these woods the lightest and consequently the most useful are the *agnus castus* and willow ... Plane has flexibility, but of a moist kind, like alder; a drier flexibility belongs to elm, ash, mulberry and cherry, but it is heavier ... (Pliny, *Hist. Nat.* 16.77)

Polybius is another valuable primary source of information on the construction of shields, describing the *scutum* being formed from at least two layers of wood laid in alternate directions (plywood), facing with fabric followed by skin (6.23.2–7). Along with Plutarch, he describes the use of metal binding to strengthen the edges and prevent the plywood springing apart (Polybius, 6.23.4; Plutarch, *Camillus*, 40.4), although this appears to have been a practice more widely used in Europe, no evidence of metal binding having been found on any of the shields at Dura. Polybius also describes the shape of the longer legionary form (1.22.5; 6.23.3), as does Cassius Dio in the third century AD, of semi-cylindrical shields being 'hollowed and channel-like' (40.30.1). This is again echoed by Ammianus, with his description of convex-shaped shields, '*patula ... et incurva*', being used as floats to cross water (24.6.7).

Ammianus also describes the convention of decorating shields with devices being used by Germans as a means to identify Roman troop units (Ammianus, 16.12.6). This practice is attested by Claudian, who describes the 'brave regiment of Leones, to whose name their

Fig 6. Cavalry *stela*. Described as Imperial copy of a Republican relief of Mettius Curtius, Lacus Curtius in the Forum, Rome. (Artwork by J. R. Travis)

shields bear witness' (Claudian, *Bel. Gild.* 423). Similarly, Vegetius records each cohort of a legion having its own distinctive shield (2.18), so clearly not only did each legion have its own individual device, but in the same way, each cohort within that legion could also be easily identified in combat situations. This could refer to decoration with applied metallic shield furniture, but most probably suggests painted decoration. It is known from the scant archaeological evidence that shields were painted to some degree, and logically the use of paint would not only have served to waterproof and protect the structure of the board, but could also have augmented its ability to withstand or absorb blows (see chapter VII).

Further, both Caesar and Cassius Dio describe the use of removable protective shield covers (again, as discussed in chapter VII), examples of which have been found made of leather from Europe, for both rectangular and oval types (Caesar, *Bell. Gall,* 2.21.5; Cassius Dio, 61.3). These would have helped to protect the shield from the elements when not in use, further prolonging its effective life. As with the shield board itself, archaeological evidence suggests that these were also decorated by the addition of appliqué panels, with animal and other motifs representative of the user's legion.

Included as an essential part of personal protection equipment for all military personnel, the shield's purpose was not entirely confined to protection, having the potential also to be used offensively, as attested by both sculptural depictions (including images on Trajan's Column) and descriptions within several reliable literary sources (discussed further in chapters VIII and IX). The *scutum* offered a substantial protective barrier for the individual legionary, the curved shape wrapping partially around the body, also providing a measure of side protection. In addition, the majority of projectiles, being aimed directly at the bearer, would, due to the curvature, in all probability strike the body of the shield at an angle and be deflected. Its efficiency is attested by Caesar in his description of how one *scutum* was able to stop 120 arrows (*Bell. Gall,* 2.21.5).

Legionary troops, however, did not by preference fight individually, many notable victories being credited to their teamwork as bodies of men, each protecting his neighbour. A range of strategic techniques are known to have been used, some offensive, others defensive. For example, Tacitus described legionaries, in close-order fighting, using the formation of the shield wall to bunch opponents into confined spaces (*Hist.* 2.22; 2.42), while Ammianus described the use of the *testudo* (6.12.44; 20.11.8; 26.6.16; Figs 69–71). In addition to the ability of the shield board to provide a defensive barrier to direct attack, to deflect projectiles and as a means of crowd control (techniques borrowed today by modern police riot squads), the shield could also be used as weapon. The substantial shield boss (*umbo*) could also be used to strike an opponent in close-quarters fighting, as described by both Tacitus and Livy (Tacitus, *Histories* 4.29; Tacitus, *Agric.* 36; Livy, 9.41.118).

SCULPTURAL SOURCES

There are a number of different forms of sculptural sources available that are sources of information on military equipment, including grave *stelae* and monumental sculptures, all of which have inherent advantages and disadvantages in the accuracy of that information.

As historians, we are always advised to have regard for potential bias within written histories, in that they are always written by the winning side; the vanquished, by necessity, are depicted with the extreme qualities of weakness, stupidity and malevolence in order to better promote the victor. Sculptural/artistic representations are to be viewed in the same light. As historical documents their content may not accurately represent reality, but nevertheless they are still a valuable resource for historical sequencing, chronology and equipment, provided that their nature is understood.

There are a number of important ceremonial sculptures that are of boundless value to the study of Roman military actions and equipment. The most notable of these is Trajan's Column. Dedicated in AD 113, the Column was erected within Trajan's monumental Forum complex at Rome (built between AD 106–113), inside a courtyard flanked by the *Basilica Ulpia* to the south side, the Greek and Latin libraries on the east and west (the *Bibliotheca Ulpia*) and Temple of the Divine Trajan on the north (Davies, 1997, 43; Osborne, 1970, 1155). The whole complex of buildings, with the Column inside, was reputedly designed by the Syrian artist/

Fig 7. Scene from Trajan's Column showing 'legionaries' in *lorica segmentata*, 'auxiliaries' in mail (*hamata*) and archers in scale (*squamata*). (From Cichorius, 1896, *Die Reliefs der Traianssäule*, plate LXXXVI)

architect, Apollodorus of Damascus, who had accompanied Trajan on the Dacian campaigns (Davies, 1997, 43).

Trajan's Column is one of the best-known examples of Roman monumental sculpture and is widely used for its well-defined images of Roman military personnel, legionary and auxiliary. Several types of shield can be seen on the Column, including both of the main types widely used: the semi-cylindrical, rectangular shield usually associated with legionary troops, and the lighter, auxiliary and cavalry shields which are mostly oval but occasionally long and hexagonal in shape (Figs 7 & 58).

However, caution must be exercised in using these images as definitive of Roman military equipment. As with the other grave *stelae* and monumental sculptural reliefs, artistic methods were employed that may mislead. Scale of equipment, particularly helmets and shields, was reduced to fit the available space and not to obscure the bearer. Furthermore, the images depicted represent a snapshot in time, dating only to the period of the Dacian wars, and cannot therefore allow any insight into earlier or later periods, nor can they show any developmental progression. The quality of the work is undoubtedly excellent, being carried out by skilled artists, although it is unlikely that they possessed any great knowledge of military equipment, other than ceremonial or processional, as would have been seen in the capital city of Rome by any other citizen. As ceremonial armour, even in modern times, has always drawn more from archaic models, the equipment depicted may be more representative of equipment from earlier periods.

The Column consists of a pedestal base (with a small chamber housing the emperor's ashes), a main column above, topped by a Tuscan capital with pedestal and statue at its apex (originally a gilded statue of the emperor Trajan holding a spear and orb, but replaced by one of St Peter). There are a number of works written describing the Column, its purpose and methods of construction, and the artistic and historical merits of its decoration. In many of these works its height fluctuates between 150 Roman feet (44.07 metres) and 100 Roman feet (around 125 feet, or 38.4 metres) to the base of the statue, including the pedestal base (Osborne, 1970, 1155; Davies, 1997, 43; Lancaster, 1999, 419–439).

Lancaster (1999, 419–439) discussed in depth the methods of construction of the Column, considering its foundations, its sequence of construction in relation to other parts of the Forum complex and the sourcing of materials, and conjectured reconstructions of the lifting frame/tower used in its erection. Other authors, however, discuss the reasons for its existence – whether it was made to mark the achievements of Trajan, or of the individual members of the army, in the campaign itself (Osborne, 1970, 1155); whether to celebrate Trajan's success (to justify the use of the Column base as a repository for his ashes; Davies, 1997, 63–65); or, as suggested by Richmond (1982, 1), to mark the technological and constructional achievement of the architects (the height of the Column representing the height of the overburden of earth removed from the hill in order to build the Forum).

The cylindrical pillar of Parian marble (formed from stone sourced from the Luna quarries, 300 km north of Rome) was provided with a central, helical stairwell. This is illuminated by small windows, barely visible within the exterior decorative spiral frieze of relief sculpture depicting Trajan's victory in the Dacian wars of AD 101–106. The relief winds around the shaft in twenty-three revolutions, 1 metre wide by 198 metres long (215 yards), which Davies (1997, 41) proposed may have been based either on the continuous illustrated scrolls (*rotulus*), or painted lengths of fabric wound around the columns of temples on feast days. It bears a pictorial documentation and commemoration of Trajan's victory in the two Dacian wars. It portrays symbolically the men who fought and died on both sides (Rossi, 1971, 14), and Richmond (1982, 1) believed that it depicted the 'best illustrations of the army' (it is undoubtedly one of the most notable pictorial representations of the Roman army, used as a 'primary' source by scholars for centuries, with many suggested 'reconstructions' in academic works and fine art based on its imagery). However, more recent opinions now agree that this image of the Roman army should be viewed with caution. While the artistic quality of the sculpture is superb, the representations of the figures depicted are highly stylized, and the equipment shown is of dubious accuracy.

Debate also surrounds the work of authors discussing the features and merits of the Column. Numerous works exist describing the Column, methods of its construction, reasons for its existence, and interpretation of featured events and equipment used, including Cichorius (1896), who produced a record of high-quality photographic images of the entire Column (from casts made in the 1860s by Napoleon III) that are still used by many modern authors (including Lepper & Frere, 1988). However, in more recent years, caveats have been voiced. For example, Bishop and Coulston, among others, refute the reliability of the depictions of equipment, which, although supposedly showing a 'snapshot' of the military in the Trajanic period, does not reflect the hard evidence of actual equipment found in the archaeological record (Bishop & Coulston, 2006, 35, 254–259).

Similarly, interpretations of the stylistic conventions employed, by authors such as Rossi (1971), have been seen by some authors as outdated. Rossi (1971, 14) suggested that the artist had utilized a convention that separated the civilian legionary from the non-civilian auxiliary visually, by depicting all of the former in short tunic and bare legs, wearing *lorica segmentata* and carrying rectangular shield and *pilum*. However, other contemporary sculptural representations suggest mail still being worn by some legionary troops at this time. For example, legionaries are depicted in *hamata* in the metopes of the *Tropaeum Traiani* at Adamklissi (Figs 8–11; Russell-Robinson, 1975, 170–171). Richmond suggested that these had been specifically provided with *hamata* to arm them against the Dacian *falx* (Richmond, 1967, 34–35; Russell-Robinson, 1975, 170). The auxiliaries, in contrast, were depicted either in mail, short trousers (*bracchae*) and with oval shield, or in regional/ethnic costume, as with the Syrian archers. Rossi also proposed the convention of using a single figure to represent a whole unit of men (not supported by Lepper & Frere, among others), the individual units being identified by their different shield designs. It is this information that Rossi (1971, 14) then used to produce his analysis of shield decoration in relation to specific units.

Many historians favour the design conventions proposed by Lepper and Frere (1988), who suggest a convention somewhat similar to the filmmaker's 'storyboard', with 'close-ups' to

Fig 8. Legionary from metopes of the *Tropaeum Traiani* at Adamklissi (metope 20), fighting Dacians with long, curved *falx* weapons. (Artwork by J. R. Travis)

Fig 9. Legionary from metopes of the *Tropaeum Traiani* at Adamklissi, fighting Dacians with long, curved *falx* weapons. (Artwork by J. R. Travis)

Fig 10. Legionary from metopes of the *Tropaeum Traiani* at Adamklissi (metope 21), fighting Dacians, wearing *squamata*, reinforced helmet, short greaves and laminated arm defence on sword arm. (Artwork by J. R. Travis)

Fig 11. Three legionaries from metopes of the *Tropaeum Traiani* at Adamklissi (metope 14). (Artwork by J. R. Travis)

focus on incidents of important action. Richmond (1982, 2) viewed the Column as a 'picture book' of individual episodes showing the everyday activities of the army. These were then converted into a unified running sequence of interlocking scenes, produced by 'working up' the wartime sketchbooks of artists travelling with the army, whom he compared to the nineteenth-century journalist-artists (forerunners to modern TV and newspaper war correspondents). Each scene of the Column could then have been based on sketches made during the campaign, which would allow for more precise details of buildings, costume, accoutrements and physiognomy of participants (Romans, auxiliaries, Dacians and allies) to be shown. In his view, any 'mistakes' would have been due to the stonemason's interpretation of these drawings (Richmond, 1982, 5).

The work of Lepper and Frere, recognised as a major study of the Column, reproduces the Cichorius plates, although disappointingly reduced in scale for reasons of economy, greatly diminishing the value of their efforts. Their work discusses the reasons behind the campaign, but is, for the main part, a 'travelogue' attempting to relate the scenes depicted to the journey to Dacia undertaken by the army. However, they give little attention to the actual equipment used, most of the text referring to the mule train rather than the troops themselves (1988, 266–269) and strongly referencing the work of Russell-Robinson (1975).

Rossi, in his slightly earlier work (1971, 14), attempted to identify the units involved in the campaign by study of their shield decorations. Lepper and Frere, in contrast, consider the shield designs to be of little significance as they thought these were left blank on the design cartoon by the artist and the 'blanks' filled in at will by the stone masons (Lepper & Frere, 1988, 31). The shield designs as identified by Rossi, along with an almost identical range of designs identified by Florescu (1965), are also reproduced in the recent work of James (2004, 165) in his discussion of possible shield designs. A composite derived from these designs is therefore reproduced here in Figs 52 to 56. As Roman shields were undoubtedly decorated (as evidenced by the archaeological remains from Dura-Europos), any discussion of their possible significance must be considered. For this reason, the works of Rossi and Florescu should not be discarded as 'outdated', to be superseded by that of Lepper and Frere (only dating from a short period later, and which concentrates on entirely different aspects of the campaign), and their work will therefore still feature in discussions of shield designs in this study.

Ultimately, any discussion of the military equipment shown on the Column should consider that the accurate depiction of the Roman army was not its primary function. As discussed earlier, Richmond (1982, 1) believed that the Column served three purposes: representing, through its height, the height of the cliff excavated to produce the Forum complex; providing a war memorial for the two Dacian campaigns (the first consisting of a series of expeditions leading up to the second campaign of organized conquest); and third, functioning as a repository for the emperor's ashes. Alternately, Davies (1997, 46–65) considered the Column to have been commissioned by Trajan as his future tomb from its conception, initially masking it as a victory monument, to be adapted after his death as a permanent tribute to his memory. With this its intended purpose, any accuracy of equipment, events or locations depicted would therefore have been a secondary concern.

It has been suggested that the Column was initially only decorated on its pedestal base, with the main shaft left blank, the spiral frieze possibly having been a later addition by Hadrian around the time when it became used, in its final phase, as Trajan's tomb (Richmond (1982, 1). However, Davies (1997, 63–65) considered that, if this had been the case, Hadrian would then have featured more prominently in the events depicted. She viewed the primary intended purpose of the Column as that of a funerary monument (the spiral motion of the frieze re-enacting ancient funerary rituals, proceeding around shaft), despite its initial construction taking place four years before Trajan's death. In AD 117, Trajan suffered a stroke while on campaign in Syria. He then died on the journey home, in Cilicia (around 8 August), was cremated and his ashes returned to Rome sealed in a golden urn, which was deposited in the chamber in the base of the Column (Davies, 1997, 41).

Davies proposed that the Column was erected during Trajan's lifetime, designed as a tomb from its first concept, but with its purpose masked as that of a victory monument in order to boost public support for the emperor and his Dacian wars (as the emperor was not entitled to burial within the *pomerium*, but could only be honoured in that fashion by command of the Senate after his death).

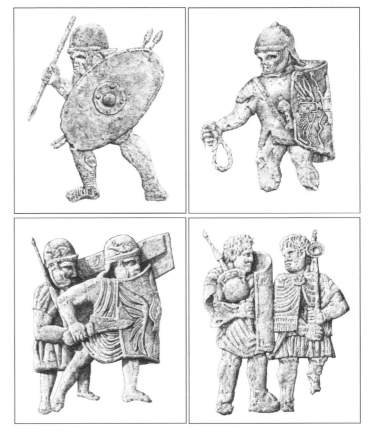

Fig 12. Figures from pedestal relief from Kaestrich fort at Mainz: (a) auxiliary soldier; (b) returning soldier; (c) fighting legionaries; (d) marching legionaries. (Artwork by J. R. Travis)

She considered that the chamber in its base had been deliberately designed to later accommodate Trajan's ashes and funerary altar, citing 'funerary' sculptural motifs elsewhere within the complex while pointing out the lack of chamber in the Marcus Aurelius Column, which had never been intended to serve as a tomb. In her study of the 'commemorative' imagery on the Column, Davies discussed the power of the dynamics of the architectural design in ensuring the perpetual memory of the emperor in the minds of the ancient visitor, which was achieved by exploiting the dramatic effect of emerging into the light at the top of the dark, helical staircase, and the subsequent vista from the summit across the Forum complex.

Fig 13. Altar of Ahenobarbus, featuring soldiers in *lorica hamata* with Celtic-style 'cape' shoulder protection, carrying oval shields with long, vertical *spina*. (Photograph by Marie-Lan Nguyen/Wikimedia Commons)

The artistic conventions used on the Column are therefore arguably of greater significance than the study of the armour itself. At first glance, the representations of the equipment carried by the men appear to be of equally high quality, but closer examination of the specific armour types, such as helmets and *lorica segmentata*, suggests that these are not as realistic as they at first appear. They do not seem to resemble any real types that have ever found. Helmets are unusually small and the buckle fastenings on the *segmentata* girdle plates would not be feasible. Representation of mail is often stylised as dots or wavy lines, some figures wearing plain tops (without the dots or wavy lines), which has led to suggestions of this being 'leather' armour. This theory was not supported by Russell-Robinson, however, who believed that the details may have been painted on instead.

Representations of shields and other military equipment can also be seen on other sculptural sources, some from earlier and some from later periods, which can be combined with the images from Trajan's Column to give a greater understanding of typology and developmental progression (Fig. 12). The Altar of Domitius Ahenobarbus (Fig. 13), dating from the second half of the first century BC, contains images of Roman legionaries, wearing Montefortino-type helmets and the long *lorica hamata* with shoulder doubling of the late Republican period (Russell-Robinson, 1975, pls 463–4). They can be seen to carry the longer, curved oval shield in use by legionary troops at that time (as described by Polybius, 1.22.5; 6.23.3), similar in form to that found at Fayum (Kimmig, 1940), with an oval *umbo* (boss) and long *spina* (central rib).

By the time the reliefs of the Mausoleum of Plancus were produced just a short time later (20–10 BC), the rectangular, semi-cylindrical legionary shield was in evidence (developed under Augustus). While this straight-sided form appears on military tombstones into the second century AD, the transitional form, with curved sides and flattened ends, also seems to have been in use, at least into the beginning of the first century AD, as can be seen on the legionary tombstone of Gnaeus Musius, from Mainz (Fig. 14; Russell-Robinson, 1975, plate 468), combined with a rectangular shield boss. However, contemporary with this, also dating from the beginning of the first century AD, Caius Valerius Crispus, of *legio VIII Augusta*, is depicted on his tombstone carrying the fully developed, rectangular, semi-cylindrical, straight-sided shield (Fig. 15; Russell-Robinson, 1975, plate 469). This has a large, rectangular boss, similar to that found at Durham (also from its inscription belonging to a member

Fig 14. Transitional shield form, with curved sides and flattened ends, still in use at the beginning of the first century AD, as seen on the legionary tombstone of Gnaeus Musius, from Mainz. (Artwork by J. R. Travis, from Russell-Robinson, 1975)

Fig 15. Caius Valerius Crispus, of *legio VIII Augusta*, from the beginning of the first century AD. (Artwork by J. R. Travis, from Russell-Robinson, 1975, 167, plate 469)

of *legio VIII Augusta*; Fig. 45), and similar to that which is proposed to have been used on the Dura shield.

Other similar stylised images exist on ceremonial monumental structures. These include sculptures of Roman legionary and cavalrymen from the Aemilius Paulus monument at Delphi, *c.* 168 BC (Russell-Robinson, 1975, 165, plate 460; Fig. 17); the Great Trajanic Frieze (contemporary with Trajan's Column) incorporated into the later Arch of Constantine in the *Forum Romanorum* (Russell-Robinson, 1975, 182, plate 494); on the columns of Antoninus Pius (Russell-Robinson, 1975, 184, plate 497); and Marcus Aurelius (Fig. 16; Russell-Robinson, 1975, 185, plate 498–9); and on the Arch of Severus (Russell-Robinson, 1975, 183, Fig. 189). Although of undisputed artistic quality, it is a possibility that these monumental artists may not have had a significant personal experience of the military. They may have only ever seen the army in ceremonial uniforms, during triumphal marches, and possibly then only from a distance. They may not have had first-hand knowledge of the specific equipment of units, which could then have been problematic for the depicted shield designs. That the shields have been found to have been decorated suggests that such decoration was commonplace, but the artist may not necessarily have assigned the correct design to the correct unit. If a modern artist were asked to draw a soldier, without benefit of a model, he may produce something soldier-ish, but not good enough to fool a fellow soldier. Similarly, these depictions of Roman soldiers may have looked convincing to the artist and the general public of Rome, but may not accurately reflect reality.

By contrast, grave *stelae* can provide detail on shield shape and decoration in relation to known military unit, function and dating context. The artistic quality of grave *stelae* is usually not as high as on monumental sculptures, but this is compensated by the level of technical knowledge of the funerary mason. Because of their very nature *stelae* will have been produced at relatively short notice, soon after the demise of the subject. They are not usually the highly decorated, well-planned, expensive works described above. They would probably have been produced by a stonemason attached to a unit, probably himself a soldier, whose

Fig 16. Schematic of a scene from the Aurelian Column, showing legionaries carrying shields with a range of shapes. (Artwork by J. R. Travis)

Fig 17. Two legionaries, wearing military cloak (*sagum*), hands resting on shields (straight top, curved sides, decorated with thunder flashes and stars). (Artwork by J. R. Travis, from Russell-Robinson, 1975)

main job would have been the production of numerous building blocks for construction projects – forts, gateways, aqueducts, and perhaps even Hadrian's Wall. While some were accomplished stonemasons, they were not necessarily 'artists' in the sculptural sense. These 'stone cutters' are described by Tarruntenus Paternus as being among the list of men within the legion known as '*immunes*', exempted from 'more onerous duties' (*Digesta: Corpus Iuris Civilis*, 50.6.7; Campbell, 2000, 30).

They would have had first-hand, comprehensive knowledge of the equipment which they were trying to depict. Their accuracy is therefore almost on a level with a photograph. However, there are still some conventions at play, which must be considered. Three main basic conventions can be seen in designs of funerary *stelae*. The deceased may be shown standing, facing forward; he may be seated, flanked by aspects depicting a banquet; or if he belonged to a cavalry unit, he may be depicted on horseback, possibly trampling an enemy underfoot, perhaps followed by a smaller figure representing his groom. Again, this latter convention is a symbolic image, as with the monumental sculptures, promoting the concept of Roman superiority. Noted cavalry *stela* include those of Longinus, AD 43–61, from Colchester (Russell-Robinson, 1975, 106, plate 306); Flavinus of the *Ala Petriana*, from Corbridge (Russell-Robinson, 1975, 106, plate 307); and that of the recently discovered cavalryman from Lancaster.

In order to fit the image neatly into the given space, often large objects, such as spear (*hasta*), javelin (*pilum*), shield or standard, may be shrunk down considerably. Other features, like mail, as seen on monumental works, being difficult to reproduce in stone, may have been painted on, with the stone surface left smooth (leading many now to speculate on existence of leather armour). Objects, such as shields, would not be depicted fully in 3D as the sculpture would not require that level of depth and so would be flattened, not possessing the true curvature of the original. Soldiers can be shown without helmets, so as not as to obscure the facial features, and to make the subject recognisable. In later periods, armour may not be worn at all, the subject being depicted wearing his off-duty stand-down kit of tunic, belt and sword (or may be seen wearing a military cloak, or *sagum*, similarly obscuring equipment, as with the legionaries depicted on Fig. 17), which is of minimal use when studying the equipment (although in the latter example, at least their shields are clearly visible).

OTHER FORMS OF VISUAL IMAGERY

There are a plethora of visual references which cover the history and development of the shield during the Roman period of 800 BC down to AD 400, in addition to those monumental sculptural examples and funerary *stelae* described above. These take the form of imagery on coins and pottery, as well as on wall paintings, mosaics, figurines (votive or otherwise) in stone, terracotta and bronze, and, towards the end of the Empire in the west, illustrations in *codices* and other documents.

Votive offerings may exhibit as miniature shields; or as miniature figurines carrying shields. In addition, there are also examples of terracotta and bronze figurines carrying shields which were not votive in function, but may have been toys, ornaments, grave goods left in tombs, or portable representations of deities. The earliest depiction of Etruscan and Latin shields date to the ninth and eighth centuries BC and take the form of clay and bronze votive offerings. In some cases they comprise of small bronze miniature shields, like that of the *ancile* (pl. *ancilia*) found at the warrior tomb in the Bologna region of Italy. This small bronze miniature (measuring 335 mm by 305 mm) from the Tombe du Bouclier is slightly oval in shape with distinctive small, deep, semi-circular cutouts, which Sekunda *et al.* (1995, 10) suggested may be the ancestor of the *ancile* type of early shield, as carried by the Salii.

Existing from the pre-Republican period and continuing down through the Imperial period, there were in Rome two colleges of 'Salii', or 'Priests of Mars', linked with the two founding communities of the Palatine and Quirinal Hills, who were associated with a particular type of shield, the *ancile*, which was traditionally oval in shape with indented sides, often described as a 'figure-of-eight' shape. A Mycenaean origin had been proposed for the shape, but Sekunda *et al.* (1995, 10) preferred a local Italian origin (based on votive forms from Picenum and adjacent regions appearing from around 700 BC), spreading across Apennines into Latium prior to adoption of the hoplite panoply and tactics. Images of this type of shield being carried by the Salii are seen on *intaglio* seals of the fourth or third century BC, and continue to appear on the reverse of bronze denarii as late as 17 BC (Sekunda *et al.*, 1995, 9–11). However, in the case of the miniature from the Tombe du Bouclier above, the form is not what you could describe as like a 'figure of eight', being almost circular rather than long and oval, and with its 'cut outs' at the top and bottom rather than at the sides.

Dating to the early seventh century BC, a small funerary sculpture from Vetulonia of an oval-shaped shield exhibits an interesting combination of identifying features. The oval shape, along with other decorative features (possible leather edging and wooden barleycorn-shaped boss), appears to show Celtic influence, although surface decoration also shows features suggesting a ply construction, which may be an Italic tradition and which then appears to continue down through the Roman Republican period into the Imperial period, as seen in the construction of the curved, rectangular legionary *scutum* (Sekunda & McBride, 1996, 7).

The remains of a round Etruscan parade shield were found from the Esquiline Tomb 94, possibly originating from the major manufacturing centre of Tarquinia. It was made of extremely thin sheet bronze, highly ornamental with repoussé decoration, any wooden base being no longer in existence. It was 61 cm in diameter, with a central handle and a series of terminals stapled to the back of the shield, presumably designed for attachment to straps (an interpretation based on evidence from artistic representations on wall paintings, sculptures and painted pottery). It was found with a 'Calotte'-type helmet, both then being dated to the first half of the seventh century BC (Sekunda *et al.*, 1995, 5–6).

Small votive figures carrying round shields, similar to this ornamental shield from Esquiline Tomb 94, are also common from this period and tend to be known as 'Villanovan', after the people who occupied northern and central Italy during the early Iron Age. For example, an Etruscan statuette of a warrior on a candelabrum from the 'Circolo del Tritone', Vetulonia, now in Florence (Sekunda *et al.*, 1995, 6), carries on his back a more robust version of this Etruscan parade shield as would have actually been used in combat, using a similar rear strapping arrangement. The warrior/figurine carries his shield on his back suspended from his shoulders, leaving his right hand free to throw his spear (now missing), holding a mace in his left hand. This practice of the shield being worn on the back, in order to free the hands for using weapons, such as maces and axes, is reminiscent of that later employed by the housecarls of the late eleventh century AD.

During the seventh century BC, many of the Italian tribes such as the Etruscans, Latins, etc. developed the Greek 'hoplite' panoply. Commencing around 675 BC in Greece, the use of Greek-style hoplite armour and shields then spread, traditionally reaching Etruria around 600 BC, later reaching to other Latins. Representations of this type of shield are found in many forms, ranging from the wall paintings in tombs to sculptures and votive offerings and small bronze figurines.

For example, fragments of a fictile (moulded) plaque from the temple of Mater Matuta in Satricum, a town in Latium, shows a warrior using a circular 'hoplite' shield, from late sixth or early fifth century BC (from Museo Nazionale di Villa Giulia, Rome; Sekunda *et al.*, 1995, 16). Similarly, a fifth-century BC terracotta from Veii shows Aeneas carrying his father Anchises (from the legendary Trojan migration to Latium), with Aeneas wearing a Latin-style Italo-Attic helmet combined with a hoplite-style shield (Sekunda *et al.*, 1995, 17), which also highlights how sometimes only parts of the hoplite equipment panoply may have been adopted, along with the hoplite-style phalanx formation fighting.

In contrast, two terracotta figurines from Veii, now in the Museo Nazionale di Villa Giulia, Rome, dating from the beginning of the fifth century BC (Sekunda *et al.*, 1995, 13), depict a pair of naked young men engaged in a war dance, the steps of which were designed to train the young warrior in the moves for single combat (similar, perhaps, can be seen in the set 'kata' of the traditional Japanese martial arts in use today). The figures are stood in identical poses, one figure 'dancing' with a round 'hoplite' shield, the other with a square shield with barleycorn-shaped central rib, indicative again, perhaps, of the Latin cities' mixed degree of take-up of the 'Greek' hoplite panoply and tactics.

This mixed panoply is also echoed within the literature references, such as Diodorus, who reports that 'in ancient times, when the Romans used rectangular shields, the Etruscans fought in phalanx using bronze shields, but having compelled the Romans to adopt the same equipment they were themselves defeated'. This also informs us that the rectangular shield had already been in use by the Romans for some time previously, and was no new introduction following the round hoplite shield ('*aspis*').

Imagery of shields shown in use by warriors using other contemporary parts of the armour assemblage can be seen on several ceramic and bronze bowls (for example the Certosa and Arnoaldi *situlae*), decorated either in relief or painted medium. The Certosa *situla* is a bronze 'vase', highly decorated with rows of figures representing Venetic warriors, those on the uppermost row showing two cavalrymen leading four groups of infantry (a group of five figures, followed by three groups of four figures), in a possible precursor to the manipular battle formation: the first group carry long oval shields; the second carry square shields with rounded edges and square boss (shoulder to hip length); the third carry round 'hoplite'-style '*aspis*' shields, with triangular edge decoration (again shoulder to hip length); and the last group are unshielded (Fig. 28; Zotti, 2006; Sekunda *et al.*, 1995, 34).

Similarly, the Arnoaldi *situla* is a seventh- or sixth-century BC bronze vase (resembling a bucket) from Bologna (again highly decorated, as with the Certosa *situla*), which depicts several armoured figures including two horsemen, one wearing a Villanovan-style ridge helmet and carrying a round 'hoplite' shield; it also depicts a number of infantrymen carrying long, rectangular shields with curved corners and spindle-shaped bosses (Stary, 1981, 290).

Imagery of shields and their users can also be seen in painted decoration on ceramic vessels: for example, on black-figure pottery, commencing from the seventh century BC and ending around 480 BC; this then being replaced by red-figure pottery from around the fifth to the fourth century BC.

On one black-figure 'krater', *c.* 575–550 BC, two 'hoplites' are depicted wearing an archaic armour panoply (including a 'bell' cuirass), carrying large, round hoplite shields on their left arms (one figure showing the shield reverse, with a central arm strap (*porpax*) and handgrip (*antilabe*) close to the edge; the other figure's shield showing its frontal decoration, with

'*Gorgoneion*' face motif), both apparently fighting over the body of 'Hippolytos' (in the Louvre, Paris; Campbell, 2012, 35). On another black-figure vessel, a '*hydria*' (water jar), of similar dating context at around 575–525 BC, two 'hoplites' are depicted fighting over a corpse, the design being almost identical in format to the former example, save for the fact that there are no visible shield decorations, and the men are naked except for helmets and greaves (also in the Louvre, Paris; Campbell, 2012, 29). On a third example, an Ionian black-figure amphora *c.* 520 BC (Martin von Wagner Museum, University of Würzburg), there is again a similarly formatted battle scene showing two hoplites, this time in the process of killing a third, central figure. They are depicted wearing short '*chiton*' tunics with a divided skirt, bronze 'bell' cuirasses, greaves and Corinthian helmets. Both men are again in near-identical poses to the previous examples, carrying large, round hoplite shields (one showing the reverse and the other showing the front, decorated on this occasion with a simple, swirling design).

Examples of red-figure vases include an Etruscan pottery bell krater from Falerii Veteres (Civita Castellana), *c.* 350–300 BC (from the Museo Nazionale di Villa Giulia, Rome). It is decorated in black and red with white highlights. It depicts images from the sack of Troy, including the figure of a warrior wearing contemporary Etruscan clothing and equipment, his Etruscan-style 'hoplite' shield visible from the reverse side, showing his arm passed through the central arm strap, and what appears to be cording running around the edge (as appears to have also been used on the 'Bomarzo shield' from the Vatican, described in greater detail in subsequent chapters). In contrast, another red-figure 'krater' vase from the first half of the fourth century BC (in Leipzig; Sekunda *et al.*, 1995, 42) features a naked Gallic horseman, carrying a straight-bladed Celtic sword and a huge oval shield, which would reach from shoulder to floor.

Funerary art, such as wall paintings or plaster reliefs in tombs, may feature images of the deceased in his armour, or as though engaging in hunting scenes. In one example, in the Giglionli Tomb, in the Latium region of Tarquinia, Italy, dating to sixth–fifth century BC, representations of round, life-sized, 'hoplite'-style shields (showing decorative motifs, some symbolic, some with animal imagery), along with other weaponry, can be seen painted on the walls of the tomb.

Other examples of funerary wall paintings may depict battle scenes. These 'battles' would usually be shown as being between Amazons and Greeks (although with combatants dressed as Etruscans in Etruscan art, and as Samnites in Samnite versions). For example, the sarcophagus of the Amazons from Tarquinia (in the Florence Archaeological Museum), dating to the late fourth to early third century BC, is a Greek marble 'trough', 1.94 metres in length, painted on four sides with battle scenes between Amazons and Greeks. Depicted figures can be seen using long oval shields, with the reverse of the shield visible, showing the user's left arm passing through a central strap, his hand gripping a cord 'handle' encircling the internal edge (again, as on the 'Bomarzo shield' from the Vatican). One warrior can be seen to be wearing linen armour (*linothorax*); the other wearing only a loosely fashioned blue tunic or cloak, armoured only in greaves and bronze helmet of Phrygian-Thracian type (Banti & Bizzari, 1974, 240).

Another tomb wall painting fragment from the Esquiline (Sekunda *et al.*, 1995, 44) appears to represent the surrender of the Samnite city to Q. Fabius Maximus Rullianus in 315 BC, and features a Samnite warrior carrying a huge oval shield (reaching from shoulder to floor), similar to that depicted on the red-figure 'krater' vase from Leipzig previously described. However, a painted fresco of the Francois Tomb from Vulci, dating to the second half of the fourth century BC, depicts a naked Etruscan (suggested to be Aule Vipinas) killing a Latin 'hoplite' dressed in short red tunic, equipped with muscle cuirass, greaves and round 'hoplite' shield (Sekunda *et al.*, 1995, 21).

Moulded representations of cavalry using 'hoplite' equipment (including round shields) are seen from Roman temples, but with similar moulds also used in other towns of Latium and southern Etruria (Fig. 18; Sekunda *et al.*, 1995, 15). However, these horsemen may not have been true 'cavalry' but acting as '*hippeis*', these being elite troops simply using the horses as a rapid means to reach the battle, where they would then dismount to fight as conventional 'hoplites'.

Fig 18. Sculpture of Roman legionary and cavalrymen, from Aemilius Paulus monument at Delphi, *circa* 168 BC. (Artwork by J. R. Travis)

Whereas the above examples of funerary art all relate to contexts of pre- or very early Republican periods (sixth to fourth centuries BC), other wall paintings are also known from the later Empire (such as those from Dura Europos, dating from the second and early third centuries AD), permitting insight into contemporary shield shapes, decoration and combat methods. For example, frescos such as the third-century AD 'Battle of Ebenezer' from Dura Europos depict soldiers carrying oval shields painted red, with the second-century AD 'Exodus' fresco from the Synagogue at Dura Europos depicting soldiers carrying a slightly differently shaped long oval shield.

Although the examples described above are of infantry soldiers, and are generally of a consistent shape for their purpose and dating context, from depictions found on mosaics and illuminated manuscripts etc. it can be seen that cavalry shields from the mid-second century AD may have used a variety of different shapes, with oval, rectangular or hexagonal shapes seen. In some cases influence of Germanic shield types can be seen, reflected in the increased use of oval, broad-oval and round shield boards. However, this regulates over a relatively short period, with both cavalry and infantry from the mid-third century AD tending to have similarly shaped large oval shield boards.

For example, a bronze roundel found in France, probably the badge of an officer dating to the third century AD, depicts legionaries from the *legio XX Valeria Victrix* and *legio II Augusta* carrying large oval shields. Similarly images on mosaics, such as those found in San Vitale, Ravenna, Italy; and 'The Great Hunt', from the Villa Imperiale del Casale (Piazza Armerina), dating to the fourth century AD; depict soldiers, some on horseback, with large, oval shields, with circular bosses.

Pictorial evidence from the fourth century AD onwards comes from a variety of sources and media, including illustrated documents, such as the *Notitia Dignitatum*, and illuminated manuscripts (*codices*), such as the Vatican *Aeneid*. The *Notitia Dignitatum* is a manuscript depicting western armour manufacturing sites (*fabricae*), which shows the types of armour and weapons made at each site, including images of oval shields (almost round) with specific emblems shown which can be related to known military units. This then links to a silver decorative dish from Geneva depicting one of the Valentian emperors (either Valentinian I 'the Great', emperor AD 364–375; Valentinian II, AD 375–392; Maximius, AD 393–423; or Valentinian III, AD 425–455), surrounded by his personal guard, who carry large oval shields, many of which have devices which closely resemble the shield emblems recorded in the *Notitia Dignitatum*.

Further imagery of Late Empire shields can be seen on illuminated manuscripts, such as the fifth-century AD Vergilius Vaticanus (Vatican, Biblioteca Apostolica, Cod. Vat. Lat. 3225). This is a manuscript containing fragments of Virgil's *Aeneid* made in Rome *c.* AD

400. The images again depict soldiers in mail shirts and coifs, carrying large, elongated oval shields with round bosses. Of similar (but just slightly later) dating context, the mosaic of the emperor Justinian (Byzantine emperor, AD 527–565) from the Basilica of San Vitale, Ravenna, Italy, depicts the emperor flanked by his generals and guards, their large oval shields by now carrying the chi-rho anagram among the designs, indicating the change in official religion.

Interestingly, despite the perceived uniformity of the Roman military (a misconception strengthened further by the stylised imagery of monumental sculptures such as Trajan's Column), with shield shapes conforming to a discernable developmental progression, for some ceremonial purposes or for officers performing a particular function, other sometimes 'archaic' shield shapes remain in use. For example, on the *stela* of a standard bearer ('*signifer*'), from Carrawburgh, Hadrian's Wall (*Procolitia*), dated to the third century AD, the officer is depicted carrying a hide-shaped shield, similar to those used in first-century BC Celtic Britain (Fig. 19). The 'memory' of these first-century BC hide-shaped shields can also be seen appearing on later period Roman

Fig 19. Standard bearer (*signifer*) from third century AD on *stela* from Carrawburgh, Hadrian's Wall (*Procolitia*), depicted carrying small, hide-shaped shield. (Artwork by J. R. Travis)

brooches, with one example of an enameled fibula from the General Post Office site, London, dated to the second century AD (Stead, 1991, 24).

In another example, a bronze votive miniature from Telamon depicts a small, round 'Greek'-style Roman cavalry shield, with barleycorn-shaped boss and central rib on front and central arm strap on the reverse (Secundus & McBride, 1997, 22). From the position of the rear strap, it is clear that the shield would have been deployed with the central rib in the horizontal position. A similar shield can be seen depicted in use on the cavalry *stela* of Mettius Curtius, Lacus Curtius in the Forum, Rome (described as an Imperial copy of a Republican relief), with the rider shown from the opposite side to the usual convention (seen from left, riding from left to right), so that the small round shield is visible, although on this occasion with a round boss decorated with a '*Gorgoneion*' face motif (Fig. 6). This small round cavalry shield type is also sometimes described as a '*popanum*', the name taken from the round, 'boss'-shaped cakes ('*popana*'), made for temple votive offerings, as seen on a small terracotta from Pergamene (Secundus & McBride, 1997, 19).

Also, Livy (44. 35. 19) reported in 310 BC the Samnite infantry using trapezoidal shields, wider at the top to better protect the user's upper body but narrower at the base so as not to impede mobility. A terracotta statuette of Minerva (from the Museo '*Sigismondo Castromediano*', Lecce) depicts the goddess with a similarly trapezoidal shield (Secunda *et al.*, 1995, 37). By the Late Republic this type of shield appears to have been out of fashion, although other examples have been found in sculpture, for example the relief of legionaries from Osuna, Spain, dated to the first century BC, both of whom can be seen carrying these long trapezoidal shields (straight sided, rounded corners, with slightly rounded top and bottom edges, but wider at the top than the bottom, sides tapering gently to base, reaching from shoulder to below knee), with an oval *umbo* like that on the shield from Fayum, a long *spina* and wide edging strip (Fig. 72; Russell-Robinson, 1975,164, Fig. 175). Similarly shaped shields can again be seen in images of named 'gladiators' on a relief from Venafro (*c.* 50 BC), which includes a fight between two figures, 'Chrestus' and 'Bassus', with the latter holding a trapezoidal *scutum*, although in this case it should be noted that gladiatorial equipment often intentionally made use of 'archaic' styles (Secunda *et al.*, 1995, 38).

ARCHAEOLOGICAL SOURCES

As shields are mostly made of perishable materials, preservation is often poor, depending on soil conditions. If unfavourable to organic materials it may only be possible to locate the shield furniture, such as central boss, handles and decorative/strengthening fixtures. If unfavourable to metals also, not even this would survive.

Discovery is also dependant on archaeological excavation preferences and chance. It is unlikely that anything will be located if an excavation is directed to a domestic site, or if non-intrusive archaeological methods are employed, as with aerial or geophysical surveys. Discovery is further influenced by chance of deposition. Accidental loss accounts for the majority of finds. However, a shield is a particularly large object and so is unlikely to be lost accidentally other than perhaps in combat situations. A shield would be a valued part of an individual's equipment and as such would not be discarded unless damaged beyond repair. If repair proved to be impossible, all reuseable parts would be stripped for the replacement shield and any non-reuseable organic/wooden parts burned as fuel. The only times when non-accidental loss or deliberate disposal would be likely would be when military sites were abandoned. In this case, everything which was no longer serviceable and too bulky to transport would be buried or burned to prevent reuse by enemy forces, often in wells or fort ditches (often by their nature waterlogged deposits), later to possibly be rediscovered by excavation, as for example at Newstead (Fig. 3; Curle, 1911, Plate XXXIV).

Despite this, most finds of shields and shield fittings have actually been the result of apparent accidental loss, with examples of complete shields from a number of sites, including Fayum, Doncaster and Dura Europos, and some possible shield fragments from Masada. In addition, finds of shield furniture, including bosses, are known from Dura and in England, including bosses from Kirkstone, Durham and Newton (the latter being used in funerary context, so possibly not being directly 'accidental'), handgrips and metallic edging strips from The Lunt (Coventry), Vindolanda, and Caerleon (Figs 4–5).

Finds of complete Roman shields are understandably rare, particularly in the colder, damper climates of Europe, due to the perishable nature of the materials used in their construction, although their metallic parts (boss, handgrip, edging strip and other scraps of decorative 'furniture') are occasionally found if soil conditions permit. The complete (or near-complete) shields found at Fayum, Doncaster (albeit heavily burnt) and Dura Europos are therefore all the more notable, particularly in the case of the latter examples that even retain much of their painted decoration (Table 1). The remains of shields from Masada, although fragmentary, are also of particular importance because of their implications towards methods of manufacture (Stiebel, 2007, 16–22). The further significance of these sites overall is that they collectively encompass the changing shape of shields throughout the period of their development, while also permitting insight into construction, decoration and use.

The following chapters will therefore discuss in greater detail the constructional details of a selection of these more notable archaeological examples, considering each within its position in the developmental progression from Republican to early and later Imperial periods. However, as the Roman shields did not suddenly appear within a vacuum in their fully developed forms, a discussion will precede considering possible origins from Bronze Age and earlier Iron Age examples, showing influences which may have derived from contact with northern or southern regions, or other Mediterranean cultures.

	Length	Width	Ply or Plank	Vert's	Horiz's	Material	Shape	Covering	Square	Round	Oval	Cu or Fe	Grip H/V	Ring Loops	Dating/context
Fayum Egypt	128 cm	63.5 cm	Ply 3 layers 9-10mm thick	6-10 cm x 2.3mm x10	2.5-5cm x 2.3mm x40 inner & outer	Birch	Long oval curved	Sheep felt			✓	Wood	H	✓	3rd to 1st C BC Celtic/Roman
Doncaster England	125 cm	64 cm	Ply 3 layers 10mm thick	11cm x 6	Inner 3.2mm outer 3.8mm approx 10cm	Oak verts alder horiz	straight sides rounded ends rectangular	Leather + layers of organic material (poss glue)		✓		Fe	V	✓	Roman/ Auxiliary Antonine
Dura Europos Tower 19	106 cm	86 cm (66cm curved)	Ply 3 layers 5-6mm thick	2-3mm approx 8cm	1-2mm approx 3cm	Plane + reinf strips	Rectangular slightly rounded ends semi-cylindrical	Fabric then leather then painted	✓			✗	H	✗	
Dura Europos Tower 2	93 cm	+62 cm	Ply scraps on leather cover approx 9mm thick if 3 layers	2-3mm	2-3mm	? Plane	Rectangular semi cylindrical	leather painted				✗	✗	✗	
Dura Europos	48 cm	24 cm	Ply frags 3 layers 7mm thick top right or bottom left	6-9cm x3mm	inner 8-9cm x 2mm outer 11,10,15,9cm wide x 1-2mm	? Plane + reinf strips (overlies fibre + gesso, but under paint)	Semi cylindrical	fibre + glue then gesso then paint				✗	✗	✗	
Dura Europos L7 by Tower19	12 cm	12 cm	Ply frag 3 layers				Semi cylindrical?	frags painted leather (dark red)				✗	✗	✗	

Table 1. Shield finds from Fayum, Doncaster and Dura Europos, from Kimmig, 1940; Buckland, 1978; & James, 2004.

II

DEVELOPMENT OF ROMAN SHIELDS

PRE-ROMAN ORIGINS AND DEVELOPMENTAL PROGRESSION

Shields were not new or exclusively Roman, their use being supported by depictions and archaeological finds from early Bronze Age contexts, with similar shapes and forms being used not only in the Italian peninsula but throughout Europe. Examples with similar features are also known from the Celtic regions, suggesting possible common origins or influences. Artistic representations of shields came in many forms, perhaps with images of warriors on wall paintings or pottery, with sculptural representations of individuals or in the form of miniature 'votive' shields. One common problem with all such artistic representations, however, is the lack of evidence for size, materials or methods of construction of the originals. For this we would ideally need to see actual shields, which only the archaeological record can provide. The same problems of preservation are common with all archaeological examples, being mostly constructed from perishable materials, but some well-preserved remains have been found from Scandinavian bog deposits. As these finds are for the most part formed from votive deposits, they consist of large numbers of items, deposited simultaneously, all therefore originating from similar dating and geographical contexts, for a brief snapshot, taking us from a 'famine to a feast' of information, although with a very limited temporal and cultural range.

This chapter will therefore draw together some of the available evidence for early shields from all areas of potential influence, to discuss the possible influences visible in the later developing Roman shield forms and constructional methods used.

DISTRIBUTION AND POSSIBLE ORIGINS OF IRON AGE SHIELDS

The oval shield with a spindle-shaped boss was a common type in use in the late Iron Age of central Europe, providing better body covering than the traditional round shield with central boss associated with the early Hallstatt period which it replaced. The board shape is an elongated oval, with an applied, pointed oval boss often described as being 'barleycorn'- or 'spindle'-shaped. Its use can be seen widely spread geographically from Britain, the Iberian Peninsula, north, central and southern Europe, and Egypt and the Levant in the Middle East, covering the period from the eighth century BC to the first century AD (Stary, 1981, 287).

Despite being constructed from mostly organic materials, actual finds from waterlogged deposits include those from Hjortspring (in Denmark), La Tène (in Switzerland) and England, with finds of shield furniture from Celtic tombs from England, Iberia, Eastern Europe, the Balkans and Italy. Artistic representations of the shields are also known from Mediterranean areas in contact with Celtic people (Iberia, Italy, Greece and Middle East), with further evidence provided by miniature 'votive' shields found in sanctuaries in Etruria and Austria. Roman literary sources for the Gallic Wars and illustrations show the Celts making use of these shields for a variety of different purposes and fighting styles, for use by infantry, cavalry and chariot fighters (Stary, 1981, 287).

Stary (1981, 288) suggested that the central rib which appears on the shields derives from animal skins used in construction, with the backbone area forming the rib. He considered that this may hold some symbolic/ritual purpose, referring particularly to some animal-skin-shaped shields to support this. However, he also acknowledged that the midrib may have a more practical purpose, as it helps to stabilise the structure of the shield and can also be used to deflect sword attacks.

The earliest examples of oval shields with spindle-shaped boss are found from the early Iron Age Villanova culture of central Italy. The evidence for these comes from illustrations dating from around eighth century BC (which Stary suggested resulted from contact with the Eastern Mediterranean areas stimulating pictorial art; 1981, 290; Fig. 20) and from miniature weapons and helmets found in grave contexts. He also proposed that a change in armaments can be seen in Etruria towards the end of eighth century BC, resulting from increased relations with the Middle East, which led to the disappearance of the oval shields in favour of the Aegean/Greek model. It is unclear what was being used in other cultures of central Italy, but Stary reported that Salian priests had a tradition for using long shields in ritual ceremonies dating back to the pre-urban phase of Rome. Furthermore, long shields remained in use for long periods, still being used by units of lightly armed warriors at the same time as the heavily armed hoplites of the fifth century BC and continuing into the change to manipular tactics.

Stary (*ibid.*) also proposed another possible area of origin in the north-east of the Apennines, citing art on the Situla Arnoaldi from Bologna (depicting a rider wearing a

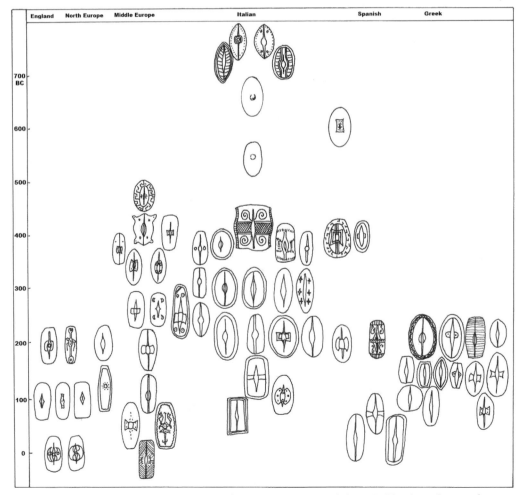

Fig 20. Evolution of the oval shield during the Iron Age. (Artwork by J. R. Travis, redrawn after Stary, 1981)

Villanovan ridge helmet and carrying an oval shield on his back) and a statuette of a similarly dressed warrior from Este (dating to around 700 BC). He further suggested that, whereas use may have ceased in Etruria from around 700 BC due to Aegean influences and adoption of Eastern Mediterranean-style armaments, in Northern Italy and Rome use continued from the early Iron Age through to the fourth century BC and into the Republican period.

In discussing the genealogy of the 'Celtic' shield, Gunby (2000, 362) also cited Eichberg (1987, 177) and Stary (1979, 183) suggesting that the oval shield, or '*scutum*', first appeared in early Iron Age Etruria, rather than the Hallstatt heartlands of central Europe ('Celtic precursors'). Citing Stary (1979, 200), she also proposed that in the eighth century BC, the Later Villanovan period, miniature oval shield replicas appear in Etruria, although by the end of the century the oval shield appears to have gone out of fashion and been replaced by the round 'Argive' shield, continuing in the Hallstatt region to the fifth century BC (Gunby, 2000, 362).

Representations from Northern Italy appearing late in the fifth century BC are often attributed to Celtic Italian disputes, but Stary (1981, 297) also noted that native Italic warriors are also represented with these shields. North of the Alps in the early Hallstatt period (HA C), Stary reported warriors still using the previously traditional round shield. However, by the early seventh century BC the earliest representations of oval shields (although transitional, still with round *umbo*) are seen on decorated cists of the Late Hallstatt period (HA D). This, he proposed, indicates a change from round to oval shields dating to around the seventh century BC in the Eastern Alps (Stary, 1981, 294).

Stary then proposed a 'Renaissance' of the oval shield type led by the Celts through their incursions into other areas, correlating with an increase in Celtic invasions in the fourth and third centuries BC in the Apennine peninsula, up to around 150 BC when the Celts were pushed back up into the Alps. This can be seen in Gaulish representations, with oval shields appearing on a number of *stelae* dating to the Celtic invasions and occupation of Northern Italy, including from Bologna, Padova, Bormio and Castiglioncello (Stary, 1981, 295).

By the fourth century BC, the Etruscans were still fighting in phalanx formation but, finding Greco-Etruscan helmets and round hoplite shields ineffective against Celtic fighting methods, they also adopted the Celtic 'Montefortino' style of helmet, with its increased visibility, along with some use of the longer shields. Similarly, during the Samnite Wars of 343 to 295 BC, Celtic alliances coincided with Samnite hoplites adopting the oval shields with spindle-shaped boss, as seen from artistic representations (Stary, 1981, 296). When faced with the lightly armed Samnite warriors and their more flexible fighting style, the Romans were then prompted to adopt similar weaponry and armour assemblage, introducing their revised 'manipular' tactics.

By the fourth century BC, the Roman army had equipped some soldiers with the '*scutum*'. Connolly (1981, 95) cited claims by Livy and Dionysius that this had been introduced for infantrymen too poor to afford the basic armour of cuirass or greaves, and was attributed to the Servian reorganisation of the Etrusco-Roman Army in the sixth century BC. However, Gunby (2000, 362) cited Stary (1981, 297), who suggested that the '*scutum*' was reintroduced to the local population by Celtic groups from central Europe moving south in the late fifth and early fourth centuries BC. She also supported Stary (1981) and Eichberg (1987, 178–181) in claiming that the '*scutum*' may have remained in use in the peripheral areas of Etruria and the Po Valley, where Greek influence was weaker. The wider readoption in the fourth century BC (along with the development of manipular formation) she saw as due to the 'demands made upon the Italic armies by the Celt's military strategies' (Gunby, 2000, 362).

Descriptions of the shape of the fifth-century BC '*scutum*' appear in both Virgil (*Aeneid*, 8.662) and Livy (*History of Rome*, 38.17), where it is named as the '*Scuta longa*'. Polybius also describes the Celtic shield ('*tureos*') as being long, oval and flat (in contrast to the circular, domed shield of the Greeks and some Italian people), and being constructed of wood, resin/glue, cloth and leather (*Histories*, 6.23). Gunby (2000, 363) noted the difference between the Roman and Celtic shield, in that the latter may not always be oval in shape (it may be oval, rectangular or hexagonal), can vary in size (it may be three-quarter length, or small, as in the Ancona frieze), and is flat. Surviving examples of shields which confirm this variability include those from La Tène, Battersea, Chertsey, Clonoura and Hjortspring.

THE HJORTSPRING SHIELDS

In 1921–22, a boat was excavated from the acidic bog at Hjortspring in Denmark, the finds published in 1937 by G. Rosenberg, which produced the 'oldest known major war booty sacrifice in Scandianavia' with 'unique evidence for iron chain mail coats and the largest number of wooden shields from a single site' (Crumlin-Pederson & Trakadas, 2003, 8). Identified as a large war canoe intended for twenty to twenty-five men, it is believed to be the only sewn-together, plank-built vessel from the pre-Roman Iron Age or earlier in Scandinavia (*ibid.*).

The Hjortspring deposit, dating to *c.* 350 BC, is over five centuries older than other votive bog deposits, such as those from Illerup Ådal, Vimose, Thorsberg and Nydam, and was made as only a single deposit (whereas the others consist of multiple deposits with varying chronology). It was found associated with a range of weapons (including spears, lances, shields and swords), therefore considered to be a military context, interpreted as booty from a conquered army (Crumlin-Pederson & Trakadas, 2003, 141). The reported finds of 'chainmail', however, consisted only of deposits of rust during excavation, which Rosenberg interpreted as remains of ten to twelve mail coats, but could alternately be a result of podzolic precipitation in an acidic bog (Crumlin-Pederson & Trakadas, 2003, 153).

At the time of deposition, the site would have appeared as a small and shallow pond, which later became a bog. The boat would have been dragged to the pond and deliberately sunk. Crumlin-Pederson and Trakadas (2003, 142) believe that the boat probably did not entirely fit and had to be pushed under by tipping on one side. As many of the weapons were found under the collapsed sides of the boat, they must therefore be considered as contemporary with its deposition. Some but not all of these weapons appear to have been deliberately destroyed prior to deposition: swords were bent, spear shafts broken, spear points bent over. The final act of the 'ceremony' appears to be that the boat was 'stoned'. The weapons found associated with the boat comprise 169 spearheads, eleven swords, several possible mail coats, and more than fifty shields. In the boat itself, finds include four iron swords and ten shields along the western side of the boat (Crumlin-Pederson & Trakadas, 2003, 142).

The wooden shields from Hjortspring form the largest collection of prehistoric shields in Europe, and exhibit a variety of sizes and shapes: some rectangular, some oval, some narrow, some broader, some slightly square; but all have smooth lateral edges and rounded corners and could be interpreted as anticipated of the typical 'Celtic' style of shield (Fig. 21). Their length varies from 61 cm to 102 cm; and their breadth varies from 29 cm to 52 cm. The largest is 88 cm by 50 cm and the smallest is 66 cm by 29 cm, although most are of an average length of 70–75 cm by a breadth of 45 cm (approximately two-thirds of the length). A smaller group of eleven or twelve are different: these are very narrow, with a breadth half of their length, slender and lightweight. Crumlin-Pederson and Trakadas (2003, 152) interpreted these as being 'suitable for more mobile warriors'.

Their construction is usually from a single piece of light, soft wood, such as alder, lime or birch. Sometimes they are formed from two or three wooden planks, held by dovetail joints, sealed by resinous putty. An oval hole was provided in the middle for the grip. This was then fastened to the sides of the hole with tenon joints. The grip was also provided with a furrow to make it more flexible.

The user's hand on the grip was then protected by addition on the exterior surface of a shield boss, carved from a single piece of wood. These bosses were oblong, diamond-shaped, protruding at the centres with raised central ridges and adjacent small furrows along the length, some with a pair of carved, round bulges at the centre. They were placed lengthwise on the central line of the shield, and fastened to the board with resin (metal was not used for the attachment; Crumlin-Pederson & Trakadas, 2003, 152).

At least sixty-four individual shields in total were identified. Many were not complete, but fifty were preserved enough to be measured to suggest their overall size. Further fragments were found of possibly up to another fifty shields, which Crumlin-Pederson and Trakadas (2003, 152) used to estimate a probable total number of about eighty, but possibly up to 100. In addition, approximately sixty-eight shield grips were found, but of these, ten then lack pegs to attach them and have traces of cord. Crumlin-Pederson & Trakadas (2003, 152) therefore estimate that only fifty-eight of these were used on shields with the remaining ten carried as spares.

Evidence of shields of similar shape to those from Hjortspring has been found elsewhere in bogs and other contexts, including those from Borremose, Kvarlov and Vaedebro.

At Borremose a leather shield cover was found. It was a rounded rectangle in shape, with a round hole in the centre for grip and boss. Around the grip hole, traces of arched ornaments were impressed into the leather. It has been C14 dated to *c.* 350 BC, and is therefore similar in date to the Hjortspring shields. At Kvarlov a wooden boss was found. It was pointed oval in shape, with a total length 21 cm by 11 cm wide, and was carved from a single piece of wood, hollowed out at the back for the user's hand. It has been C14 dated to *c.* 780–400 BC, and so is possibly older than the Hjortspring

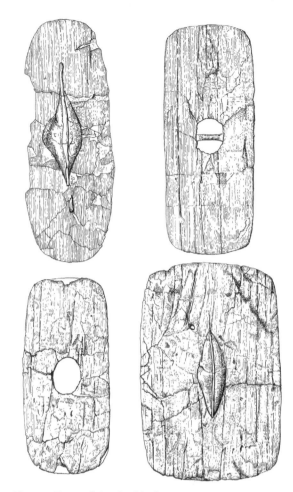

Fig 21. Four of the shields from Hjortspring, showing a range of shapes and sizes. (Artwork by J. R. Travis, redrawn after Crumlin-Pederson & Trakadas, 2003)

deposits. The boss was shaped to provide a lengthwise long, narrow rib on the outside surface. The walls of the boss were wider at the base to allow a larger area to glue to the shield. No trace of any glue remains but two wooden nails can be seen to have been driven through each side of the rib at each end. There are no traces of any applied metal, or evidence of the grip. At Vaedebro, near Skanderborg, eastern Jutland, not far from Illerup Ådal, an almost complete shield was found (along with a couple of spearheads and a broken arrow shaft) dating to the first century BC, so being somewhat later in date than those from Hjortspring bog. The shield boss was missing but the grip remained, reinforced by a piece of iron. The shield face also has traces of red paint and rows of small holes (Crumlin-Pederson & Trakadas, 2003, 171).

The Hjortspring shields are of a shape common to the European type. It is also known to have been used by the Celts and Northern Germanic cultures, but after 3000 BC these introduce the use of metal strengthening bands across the boss and centre of the shield. Those from Hjortspring in contrast have no metal used. This may be due to metal poverty or may reflect changing fashion. Crumlin-Pederson & Trakadas (2003, 172) drew comparisons to the shield found at Fayum in Egypt, noting the different method of construction (ply), the curved sides and central ridge (*spina*) on the latter, but the use of a similarly shaped boss. The Pergamon relief also shows oval or rounded, oblong shields (Pergamon

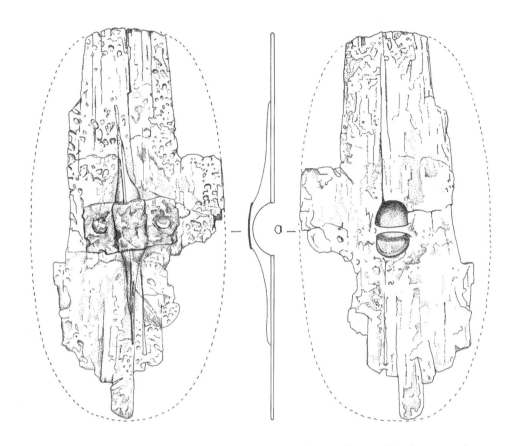

Fig 22. Celtic shields. La Tène. (Artwork by J. R. Travis, redrawn after Buckland, 1978 and Connolly, 1978)

is a Greek triumphal shrine of a battle of Greeks and Celts in Western Asia Minor). Some of the shields depicted have heavy fittings over pointed oval bosses. Crumlin-Pederson & Trakadas (2003, 172) suggested that these are later forms, dating from battles around 235 BC and 187–185 BC. Some 'Celtic'-style wooden shields, with heavy iron fittings across the bosses, have also been found from La Tène in Switzerland (one of these dating to 229 BC; Fig. 22). Crumlin-Pederson & Trakadas (2003, 172) suggested that Hjortspring shields are earlier representations of the two previous examples (Fayum and Pergamon), illustrating the distribution and level of uniformity in the 'Celtic world'.

They further suggested a chronology of Iron Age shield development with, firstly, up to 300 BC, shields being of the Hjortspring type: with oval boss but no metal fittings. From that time rectangular metal fittings over the wooden boss were introduced (between early and middle La Tène periods of *c.* 300–250 BC), with an example cited from Gournay-sur-Aronde in Northern France. By the first century BC, shields began to be fitted with circular iron bosses (*ibid.*). The Hjortspring shields then fit into the earlier phase of this chronology, through being dated by C14 to fourth or third century BC (400–250 BC) and from comparison to the rest of the finds, which have been estimated at fourth century BC.

THE CHERTSEY SHIELD

A complete example of an oval 'Celtic' bronze shield was found at Chertsey in Britain in December 1985 (Stead, 1991, 5; Fig. 23a). It was found by workmen clearing gravel from

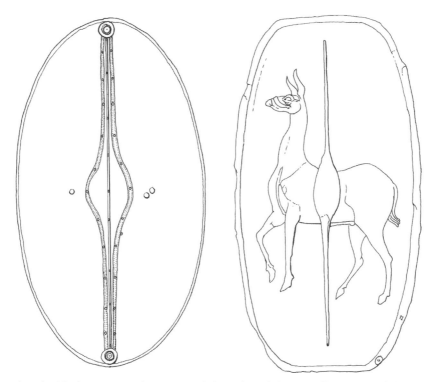

Fig 23. Celtic shields: bronze (a) Chertsey (oval shaped) and (b) Po Valley. (Artwork by J. R. Travis, redrawn after Stead, 1991, 5; & Gunby, 2000)

near to the edge of a 3–4.5-metre-deep waterlogged pit, in an old riverbed 700 metres west of the present course of the River Thames. The shield was brought up by bucket, initially missing its handle, although after a search this was later found in previously excavated material. Through personal library research it was identified by the non-archaeologist finder, who then offered the find to the British Museum.

The shield is 83.6 cm long by 46.8 cm wide, weighing 2.75 kg (Stead, 1991, 5). It is the only known Celtic shield formed entirely of bronze. It was constructed from several pieces of bronze, riveted together. In the centre it has a spindle-shaped boss, with a vertical spine made from a single length of bronze sheet. The spine is formed with a flange 8 mm wide on each side and rises to its highest point in the centre, 37 mm above the flat surface of the shield. The spine then terminates 32.5 mm from the ends of the shield, with the flanges merging into broad terminals. On the front, at each end, there is a decorative bronze disk formed in three parts with a central rivet. Two large side panels, of a consistent thickness (average 1.16 mm), then form the main shield oval shape, which are riveted to the flanges of the spine by 22 rivets on each side. A masking strip 9 mm wide then covers the join between the side panels and the central flange, further decorated with two rows of diamond shapes. This strip covers half of the above joining rivets, with every other alternate rivet passing through all three layers of masking strip, side panel and spinal flange. Along the top of the *umbo* (boss) and *spina* (spine), a 5 mm-wide semi-tubular rib is attached by nine rivets, and in the centre of the boss an oval perforation of 6 mm by 3 mm would have served for a further unknown decorative/functional attachment, now missing.

Separate edge binding is provided in a single, un-jointed piece, 16 mm wide, flat against the back surface and edged with a cordon 6–7 mm wide at the front, raised to a single ridge. Stead (1991, 5) suggested that this was constructed as a broad, oval frame attached to the rear of the shield by six rivets (one at each terminal, one through the panel at each side, and two through the panel on the lower right-hand side). Once attached, the outer edge of the frame was then turned back on itself to overlap and enclose the panel edges.

The 9.5 mm gap across the rear of the concave boss is spanned on the back of the shield by its tubular grip/handle, 106 mm long by 12–13 mm diameter, formed from a wooden core of ash (*fraxinius sp.*) covered in sheet bronze. Radiocarbon dating tests on the handle core returned a date of 2300 BP +/- 80, which when recalibrated suggests a date of 400–250 BC (68 per cent confidence) or 750–190 BC (95 per cent confidence). The handle then terminates in elaborate semi-tubular terminals resembling opposed swan's heads, attached by three rivets at each end.

Stead (1991, 6) reported that the handle shows signs of repair, by way of a tear in the repoussé decoration, and reattachment, by use of an additional rivet. This indicates that the shield was not new when deposited (sufficiently so as to cause minimum wear, but without any heavy combat damage). However, the crumpled condition on discovery and damage causing the detachment of the handle Stead (1991, 10) interpreted as relating to time of discovery rather than to any deliberate, ritual 'killing' of the artifact prior to deposition. However, he also suggested that the shield was unlikely to have been used for combat, considering that leather and wooden versions would serve better, and was more probably intended for display/ritual use (1991, 23). The position of the handle is described as being not central to the boss, which Stead believed would allow the user to more comfortably grip from above, hand pointed downwards, leaving the right hand free for a weapon (citing the viewpoint of Brunaux & Rapin, 1988). There is no evidence for the shield also being strapped to the arm in use, although a 'staple' found on the rear was interpreted as being one remaining of a pair for a shoulder carrying strap (Stead, 1991, 23).

In his discussion of the shield's possible dating and origins, Stead (1991, 21–22) related its shape (oval with spindle-shaped boss) to the typical Celtic shield as seen in Greek and Roman art (for example from Pergamon, Arch of Orange, etc.). He cited Stary (1981) for identification of the oval-shaped shield as a basic Celtic form used for centuries, seen from eighth century BC in Italy, although first seen north of the Alps in fifth-century BC Hallstatt scabbard (Pauli, 1978, 246). Stead (1991, 22) proposed that it is the shape and form of the spindle-shaped bosses and *umbo* coverings which are more diagnostic, appearing in different parts of Europe at different times, with the band-shaped *umbo* being introduced on the continent in early La Tène II contexts, but not appearing in Britain before the first century BC, with post-conquest 'barbarians' being depicted on sculptures still using simple oval shields with spindle-shaped boss. However, Stead appeared to prefer a continental origin for the Chertsey shield, as in his opinion many continental shields used metal strip covering over handles, whereas examples from Britain (with the exception of that from Battersea) were usually formed from all wood alone. This, he suggested, would allow the C14 date of 400–190

Fig 24. Gallic warrior from Vacheres, late first century BC, wearing *lorica hamata* with shoulder doubling, leaning on Celtic-style oval shield: (a) front view; (b) side view. (Artwork by J. R. Travis)

BC for the handle, to fit in with attributing to La Tène I, and so being contemporary to the Hjortspring ship (dated to 350–300 cal BC; Jensen, 1989), but being earlier than a shield from La Tène (dated to 229 BC by dendrochronology; de Navarro, 1972, 128).

Stead (1991, 23), did however, concede that not all oval shields have spindle-shaped bosses (citing for examples the round boss on the shield of the Vachères warrior, and as seen on the Arc d'Orange; Figs 24 & 59), and that not all spindle-shaped bosses are exclusively used on oval shields, being also found on long, narrow/rectangular shields (as for example from Arnoaldi; and Witham; Fig. 31); on circular shields (for example from Telamon; and Bormio); and on hide-shaped shields (from Dürrnberg bei Hallein; and Bormio).

HIDE-SHAPED 'CELTIC' SHIELDS FROM BRITAIN

In July 1988, a collection of twenty-two miniature shields, ranging in size from 44 mm to 103 mm high (known as the 'Salisbury' collection), were offered for sale by a London dealer in antiquities and later bought by the British Museum. Their provenance was somewhat shady but believed to be part of a 'hoard' found by metal detector in the West Country, possibly near Salisbury (Stead, 1991, 10). Despite the lack of datable provenance, the significance of these miniature shields comes from the information provided on handles, bindings, bosses and board shapes from apparently contemporaneous representations. That the majority of the shields are hide-shaped, with slightly bowed sides, concave ends and projecting corners, suggests that not all shields in use were the 'typical' Celtic oval shape, with the hide shape possibly being the most common form seen in Iron Age Britain (Figs 25 & 26).

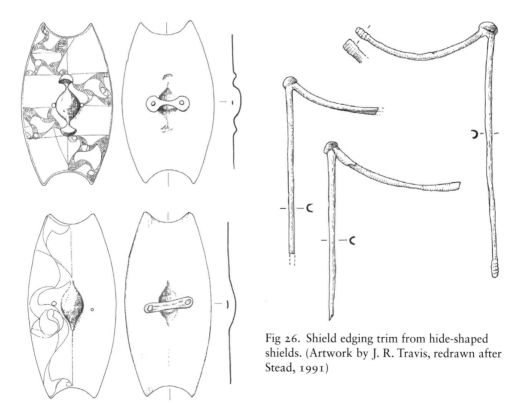

Fig 26. Shield edging trim from hide-shaped shields. (Artwork by J. R. Travis, redrawn after Stead, 1991)

Fig 25. Celtic shields: hide-shaped. (Artwork by J. R. Travis, redrawn after Stead, 1991)

On three of the shields the bosses were round, whereas the majority of others were spindle-shaped or pointed-oval-shaped, further suggesting that the 'Celtic' spindle-shaped boss was not the only form in use. Almost all of the shields indicate separate handles attached by two rivets. Most of these are just simple straps, but one was shaped to suggest a wooden core (similar to the Chertsey shield). Two handles were grooved to suggest jointing on the outside, and one had its ends inserted into slots at the side of the boss. Of the shields which exhibit some form of binding, one has separate binding strip, one has turned-over edging and four are outlined to suggest some form of binding (Stead, 1991, 11). Similarly shaped pieces of semi-tubular copper alloy binding (in lengths forming a straight length and a curved length, meeting at a projecting hollow knob), which appear to have come from these forms of full-sized shields, have been found at thirteen sites in southern England, including Spettisbury (initially identified by Piggott as part of a scabbard; Piggott, 1950, 22), Hayling Island, Deal and Wilsford Down (Stead, 1991, 20–24). Thickness of the bindings suggests use on shields which would have been 5 mm thick on the edges, with shape and size suggesting that a complete shield would require four such pieces covering the corners, with either a short gap or separate covering piece in the middle of each side and end (Stead, 1991, 21). The 'hide' shape of the shields would possibly suggest the use of hardened leather in their construction, although this may be a residual memory of much earlier shields, with the use of leather over a wooden base perhaps being more likely in view of the suggested board thickness above.

Five of the miniature shields were also found to have been decorated with asymmetrical patterns of 'Celtic'-style swirly shapes, breaking up the area into patches. These patches were then alternately left as blank voids or infilled with basketry hatching (Stead, 1991, 11). This decoration was probably imitating painted or tooled designs seen on full-sized equivalents, with stone carvings of continental 'Celtic' shields being similarly decorated, although the asymmetrical aspects of the designs appear to be a regional variant to Britain (Stead, 1991, 25).

Although a dating context is not possible for the miniature shields, due to their lack of provenance, dating is possible for some of the full-sized shield bindings found. For example, those from Deal have been attributed to second-century BC; bindings from Spettisbury dated to a 'massacre' mass grave context no later than first century BC; those from the hill fort context at Bredon Hill possibly dated to mid-first-century AD massacre level (although Stead believed this to be incorrectly dated too late; 1991, 24); Danebury context unlikely to be later than 100 BC; and those from Snettisham dating to first century BC, probably pre-Caesarian invasion. Stead (1991, 24) further suggested that all remaining finds would also be consistent with dating to first century BC, although the 'memory' of hide-shaped shields appears on Roman brooches (one example being an enameled fibula from the General Post Office site, London, dated to second century AD) suggesting the shape being still in use against the Romans well into the Imperial period (*ibid.*).

'CELTIC' IDENTIFICATION OF OVAL SHIELDS

Gunby (2000, 359–365) discussed third- and second-century BC artistic representations from the Black Sea area and the Bosphorus (from frescos, coins, tombstones and terra-cotta figures), to consider whether the presence of oval shields should be interpreted under the traditional viewpoint to indicate the possible presence of Celtic groups. These, it had been proposed, may reflect visiting Celtic mercenaries, as had also been reported by both Pausanius (1.7.2) and Polybius (5.65.10), who recorded third-century BC Celtic mercenaries serving the Ptolemaic court.

In her discussion Gunby (2000, 362) instead suggested that the identification as 'Celtic' of these representations from the Black Sea and Bosphorus may possibly be erroneous (a 'persistent misidentification'), suggesting that not all oval shields are 'Celtic' and not all 'Celtic' shields are oval.

Polybius explained that, although of similar dimensions, the Roman example is curved rather than flat. Livy (8.8) records all three classes of legionary using the shield in the Latin

Wars and as seen on Altar of Domitius Ahenobarbus. Polybius (Histories, 6.23) described its construction as being of two layers of wood, fastened with bull's hide glue, the outer surface covered with canvas then calfskin, upper and lower edges bound with iron, and an iron boss in the centre.

The '*scutum*' had reached the Greek armies by the time of the battle of Pyrrhos and the Punic Wars, then spread rapidly through the Mediterranean. Its spread was aided by mercenaries, some Galatian or Celtic. Eichberg (1987, 188–199) reported that by 281 BC the oval and rectangular '*scutum*' was in use by mercenaries in Hellenistic armies. For example, Alexandrian *stela* from the end of third century BC show large, oval *scuta* used by mercenaries (some possibly Galatians). In the second half of the third century BC, the oval shields had spread to replace Greek round shields in the Bosphorus (Triester, 1985, 133). The *scutum* also appears on coins and *Nymphaeum* graffito of an Egyptian ship, as symbols of the Ptolemaic kingdom (reign of Ptolemy II). By the first century AD they were being used from the Iberian Peninsula and Britain in the west, through northern, central and southern Europe, to the Levant and Egypt (Gunby, 2000, 363; Stary, 1981, 287).

Gunby (2000, 364) concluded that oval shields were not used exclusively by the Celts. Nor were oval shields the only type of *scutum* to have been used by the Celts since fourth century BC. Furthermore, by the third and second centuries BC, oval shields with an elongated central rib were widely distributed among non-Celtic people, being found in Greece, Asia Minor, Egypt and Rome.

Current misidentification of the *scutum* as a Celtic innovation therefore, despite contradictory archaeological and historical evidence, Gunby (2000, 364) attributed to being an 'accident of ancient history'. She believed that the ancient authors (Polybius, Livy and others) mistakenly viewed the shield as being reintroduced by the Celts in fourth century BC from beyond the Alps, whereas it should be seen as the 'Villanovan' shield. Furthermore, that depictions of Celts with oval shields with elongated rib on ancient sculptures (for example on Etruscan grave reliefs, such as the Ancona frieze and Pergamene dedication) have 'skewed perceptions' of the Celtic assemblage. Instead she believed that from the third century BC, the 'classically Celtic' large, oval *scutum* was being widely used by non-Celtic groups, whereas the true Celtic shields could vary in shape. Gunby (2000, 364) therefore disputes the identification of all figures in artistic representations as being Celtic, as other groups had used classic 'Celtic' shields and as the Celts themselves often adopted indigenous styles of dress, social organisation and ritual practice, merging their own styles with those of the local population (for example in Po Valley and in Galatia).

SUMMARY OF DEVELOPMENT

Bronze Age
The artistic and sculptural evidence available suggests that Bronze Age shields were predominately small and circular, although changing weapons, methods of warfare and organisation of armies led to some innovations of style and usage echoing that of the later Roman armies.

Fig 27. The Vulture *stela* (*circa* 2500 BC). (Artwork by J. R. Travis, from Fagan, 2004, 185).

For example, the Vulture *stela* from Telloh of around 2500 BC depicts armoured soldiers of Eannatum of Lagash, wearing simple bowl helmets and carrying spears and rectangular shields, forming a shield wall (Fig. 27; Fagan, 2004, 85). A short time later, in the sixteenth century BC, large, rectangular tower shields are shown depicted on the inlaid blade of a bronze dagger from Shaft Grave iv, from Mycenae (Buckland, 1978, 262–3; Snodgrass, 1967), a style also described by Homer, with no handgrip, carried by a short strap around the neck and under one arm (*Iliad*, 6.118).

Iron Age

Changes in warfare again by the seventh century BC led to a return in regions of Hellenic influence to armies of 'hoplites', fighting in phalanx formation using a circular shield with a central rear grip, known as a '*hoplon*' (Figs 29 & 30). The term 'Hellenic' is here used rather than that of 'Hellenistic', as the former refers to all Greek-speaking societies, whereas the latter purely refers to the style of art and culture of those societies (Howatson, 1997, 528). With cultural influences from these Hellenic regions, hoplite-style armies with Hellenic-style equipment are seen diffusing into the Italian peninsula through contact with Etruscan and coastal peoples, which is evidenced by the end of fifth century BC, on the bronze Certosa Situla depicting North Italian Arnoaldi marching spearmen (Fig. 28). Their shields Buckland (1978, 262–3; Lucke, 1962) reported as being elongated and rectangular, approximately the same shape and size as the much later Doncaster shield. However, on inspection these appear to be more rounded oval shapes, most provided with long, oval bosses running into a broad, short *spinae*. Some variation is visible among the equipment carried, however, indicating different ranks or classifications of warrior, and with one shield at least appearing to use a rectangular boss. At this time, by contrast, the Samnite people were reportedly already using the longer oval *scutum*, combined with lighter javelins and open-formation fighting techniques, which eventually influenced the adoption of the manipular legion in the fourth century BC (Livy, 4.59).

   In his discussion on the development and distribution of the typical 'Celtic' shield, Stary (1981, 297), saw the origin of oval shields with spindle-shaped boss in the early Iron Age Villanova culture of the Apennine peninsula. These then went out of fashion in Etruria at the end of the eighth century BC due to military influences from the Eastern Mediterranean, but in peripheral regions their use continued, particularly north of the Apennines. The Celts then adopted their use through cultural contacts in northern Italy and the Alps. They then brought them back with them during the invasions of Italy (fourth to third century BC), introducing them to their Samnite allies and being adopted in response by the Romans (along with the adoption of manipular formation) as being superior against more 'unpredictable' enemy tactics. Etruria also adopted some of the Celtic assemblage (particularly the increased-visibility

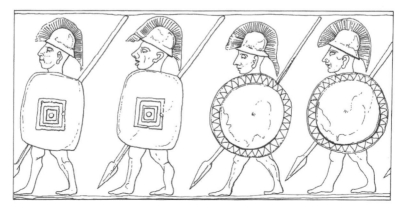

Fig 28. Warriors carrying a variety of shields (round, oval and sub-rectangular), as depicted on the Certosa *situla*. (Artwork by J. R. Travis, from Zotti, 2006)

helmet) and partial use of the oval shield for some lightly armed warriors, while still retaining their hoplite phalanx formation (Stary, 1981, 297).

By the early Iron Age in Northern Europe it is probable that the majority of shields would have been made of perishable materials (wood, wicker, leather), and as consequence little appears in the archaeological record, unless deposited perhaps in waterlogged, anaerobic conditions favourable to organic preservation. In Denmark, for example (in the Hjortspring Bog), around 100 wooden shields and bosses in were found in a presumed 'votive' deposit. These

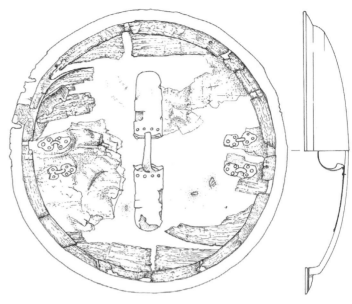

Fig 29. Remains of an Etruscan hoplite shield, from the Vatican Museum. (Artwork by J. R. Travis)

were elongated oval in shape, with convex edges, rounded corners, and had been made from a single plank of alder or lime, up to 18 mm thick at centre and 5 mm at edge. An aperture was made in this for the hand, with a separate applied horizontal grip on the rear, usually of made of ash, covered on the front by an elongated and tapering wooden boss, glued in place with a resin glue (Buckland, 1978, 262–3; Rosenberg, 1937; Crumlin-Pedersen & Trakadas, 2003).

By the later Iron Age, in areas occupied by people of the La Tène culture, similar plank-built shields are found, made of oak, one of which has been dendrochronologically dated to 256 BC (Buckland, 1978, 262–3; de Navarro, 1972). There is also by this time, from Simris in Scandinavia, some evidence of innovations such as iron or bronze edging to strengthen against vertical splitting (Buckland, 1978, 262–3; Stjernquist, 1955).

Fig 30. Statuette of an ancient Greek hoplite or heavy foot soldier, from the Berlin Museum, *circa* 1900. (Artwork by J. R. Travis)

III

REPUBLICAN PERIOD SHIELDS

In Western Europe, during the pre-Roman Iron Age and by the early Roman period, flat, oblong shields with rounded ends can be seen, both in sculptures – for example at Mondragon, Vaucluse and Mosel, near Koblenz (Buckland, 1978, 262–3) – and also in the archaeological record, with examples from Horath (Kimmig, 1940), Battersea (Fig. 31a; Buckland, 1978, 262–3; Fox, 1958, pl 14a) and Witham (Fig. 31b; Buckland, 1978, 262–3; Jope, 1971, 61–69). Further regional or changing weapon-driven developments can also be seen in later periods, with the introduction of strong metal covers over the boss; at first as just flat sheet, bent into an omega shape, to cover the long wooden *umbo*, but later with rounded hemispherical bosses appearing, particularly in Germanic regions. One such circular boss can be seen, still used in association with a long narrow *spina*, on a miniature bronze shield from Baratela, Este (Buckland, 1978, 262–3; Klindt-Jensen, 1949, Fig. 89).

Roman shields, however, differ from their Iron Age and Celtic counterparts in shape (the latter being flat, whereas the former follow the tradition residual from the Greek '*hoplon*' for curvature better able to encompass the user's body), and material, developments possibly in part being in answer to different enemies and fighting styles, and also perhaps being indicative of an advanced knowledge of material properties. Further developmental progression can then be seen over time in shape, design and method of construction from the Republican period, through the Imperial period, to the end of Empire, as seen on examples of late Roman-style oval/round plank-built shields also found preserved in sacrificial bog deposits from Illerup in Denmark (Ilkjaer, 2002, 99; Jørgensen *et al.*, 2003, 313; Fig. 42), as will be discussed in subsequent chapters.

Gunby (2000, 363) cited the Fayum shield (three layers of wooden strips, covered in a layer of felt and convex in shape) as resembling Polybius' description of the '*scutum*' (Polybius, *Histories*, 6.23). Its shape is more of a rounded rectangle, rather than oval, but she cites Diodorus Sicules and Dionysius that this is still a typical form for a Republican shield. She also cites, however, the view of Connolly (1981, 95–96, 120) that oval and rectangular forms also existed (Gunby, 2000, 363). Therefore, despite the disputed provenance for the Fayum shield, it will be discussed here below as being representative of the Roman Republican shield

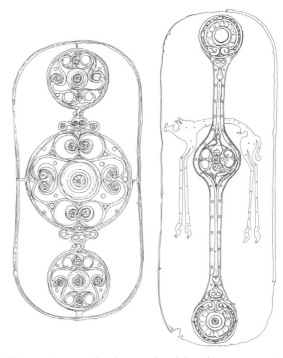

Fig 31. Rectangular, bronze-faced fourth-century BC 'La Tène' or 'Celtic' shields from Britain: (a) Battersea shield; (b) Witham shield, both currently in the British Museum. (Artwork by J. R. Travis)

because of how its features (of shape and construction methods) meet those anticipated of Roman shields of that period.

As discussed previously in chapter I (when discussing visual imagery), and in appendix I (on the origins and development of the Roman army), the armies of the Roman and other Italic people had initially used hoplite-style equipment and phalanx formation, as can be evidenced from the literary, sculptural and other visual resources. However, archaeological evidence of actual shield remains is scarce, with the notable exception of the 'Bomarzo', or 'Vatican' shield, which will therefore also be discussed below as an example of the earliest-period Roman shields.

THE 'BOMARZO' OR 'VATICAN' SHIELD

The 'Bomarzo' or 'Vatican' shield in Museo Gregoriano Etrusco (Vatican), was discovered in 1830, probably from an Etruscan tomb near Bomarzo (the site of the Etruscan city of *Volsinii*). It is one of the few remaining examples of a 'hoplite' shield (*aspis*), which is sufficiently preserved to allow us to see methods of construction (although an Etruscan version; Fig. 29; Campbell, 2012, 22). The core of the shield was a wooden bowl with flat projecting rim, made up of five separate planks (of poplar), the planking arranged so that the grain/direction of planks was horizontal in use. The flat rim was then further strengthened by the addition of a series of shaped wooden laths, with triangular cross-section, which graduated and smoothed the edge profile, on the transition from 'bowl' to 'rim', these laths being arranged in such a way that they lay across the junctions of the planks, so consolidating the structure. The thickness of the wood varied from 10–11 mm in the centre to 12–18 mm at the edges. The overall diameter of the completed shield was then approximately 82 cm (so larger than the Esquiline Tomb 94 parade shield, at only 61 cm), perhaps reflecting their functional differences: one as a 'parade' shield, the other as a 'combat' shield.

The interior surface was then covered with a thin layer of leather glued in place, and the exterior surface covered with a layer of sheet bronze (*chalkoma*), 0.5 mm thick, glued in place using pitch (Kagan & Viggiano, 2013, 157–159). This bronze surface could then have been decorated by a repouseé pattern (as on the round Etruscan parade shield from the Esquiline Tomb 94), although on other shields this has also been highly polished or painted (various shield emblems and designs are known from wall paintings and painted pottery examples previously described).

In the centre of the reverse surface, a bronze central armband (*porpax*) was riveted in place and two pairs of staples fixed, through which braided cord handgrips (*antilabe*) were attached, along with a series of additional terminals through which a spare cord was threaded (passing around the perimeter, permitting the user to quickly and easily rotate the shield by moving his hand around the cord to its next position, or to spin a full 180°). This perimeter cord could also then be used to suspend the shield over the user's shoulder when not in use, or to free up both of his hands for use of mace and axe in combat situations.

Krentz (1985, 53) discussed the function of the central armband (*porpax*), suggesting that it was an innovation to improve manoeuvrability and performance of the larger, two-handled hoplite shield (in comparison to the earlier round shields), in that it helped to better support the weight. It would allow the user to better angle the shield to deflect blows, and would not be dropped if the user lost hold of the handgrip (*antilabe*), and leaving his left hand free to carry a spare weapon. It had a minor disadvantage in that it would better protect the left side, rather than the right side or back, but this was offset by the increased manoeuvrability, lending itself better also to individual fighting and looser formations.

THE FAYUM SHIELD (REPUBLICAN PERIOD, 280S TO 100 BC)

The Archaeological Evidence

The Fayum shield was found by English papyrus hunters in 1900, at Kasr el Harit, Egypt. However it was not until 1940 that W. Kimmig (1940, 106–111), who was not the original excavator, published it. The shield is of a long, oval shape, its side edges curving inwards, built of three layers of birch ply, with a long, vertical *spina*, oval wooden *umbo* and horizontal handle to the rear (Fig. 32).

Kimmig suggested a Celtic/Galatian origin because of what he believed to be its 'Celtic' shape, based on comparison to European examples of Celtic shields, including one from Battersea (Fig. 30). His dating of the site comes from late Ptolemaic buildings (late second-century BC pottery, lamps and papyrus fragments) cut through by later Ptolemaic-style graves. It is not clear, however, from Kimmig's account, whether the shield comes from the buildings or grave contexts. Therefore, this late second-century BC '*terminus post quem*' gives us the oldest possible dating for the shield.

Initially Kimmig suggested a date range for the shield from either of two Celtic/Galatian incursions into Egypt, between the 279 BC raid on Delphi to around 100 BC. He cited Pausanias who reported 4,000 Celtic mercenaries fighting on the Egyptian side during the first Syrian War in 274 BC, between Magas of Kyrene and Ptolemy (Pausanias 1.7.2). He also referenced Polybius for a second incursion in 216 BC, with Galatian auxiliaries again fighting on the Egyptian side in the war between Antiochus III and Ptolemy IV (Polybius 5.65.10). Finally, with no additional reasoning, he opted for assigning the shield to the third century BC. However, from his report, the buildings had been dated to late second century

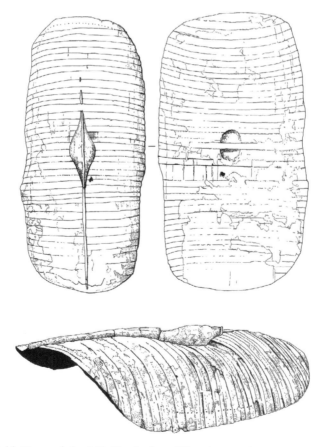

Fig 32. Fayum shield. (Artwork by J. R. Travis, from Kimmig, 1940)

BC. Therefore, if the shield dates to the buildings it may belong possibly to 150–100 BC. The buildings had then been cut through by late Ptolemaic-style graves, the shield possibly belonging to this same dated context, which therefore could even be as late as mid-first century BC.

Buckland, following Kimmig, attributed the Fayum shield to a Celtic/Galatian mercenary origin, drawing on European traditions, its construction being from birch wood, this species not being found south of Anatolia (Buckland, 1978, 262). He compared the spindle-shaped *umbo* and *spina* to elongated oval shields depicted on Pergamon weapon reliefs of the early second century BC, while acknowledging that 'the forms were not closely matched' and 'neither were exclusively Celtic', suggesting that its curved form need not be specifically Celtic, the use of plywood shields possibly being more general than appreciated, drawing comparisons to tall Roman legionary *scuta* on sculptural representations.

The Fayum Shield Board Construction

The Fayum shield has a long, oval shape, with a maximum length of 1.28 metres by a maximum width of 0.635 metres. It is constructed of layers of birch wood ply. The centre/core is formed by nine to ten vertical planks, between 6 cm and 10 cm wide, and the front and back outer surfaces formed by forty horizontal strips. These are between 2.5 cm and 5 cm wide, with the exception of the rounded ends, which are broader to protect against splintering. The overall depth of the ply is less than 1 cm thick for the three layers, with an average of 2 mm to 3 mm each layer. The shield had been covered on both sides by a thin layer of sheep felt. The exterior surface was applied first, covering the shield board up to the edge. The inside surface was then applied, overlapping by 5 cm to 6 cm. This was then stitched down using a double linen thread, the resulting seam then folded over and glued.

The *umbo* and vertical-bracing *spina* were worked in three parts. The central boss (*umbo*) is 35.5 cm long by 11.5 cm wide, carved from a solid piece of wood, hollowed out to a 6 mm-thick wall, shaped top and bottom to the junctions with the vertical *spinae*. It is fixed to the shield by four strong iron nails, hammered through the plywood into the solid part of the boss. The *spinae* were 45.5 cm long and 0.8–1.0 cm wide (although now missing from one side) and were fixed to the shield from the outside by four wooden nails each side, 10 to 12 cm apart. The 2.9 cm-wide horizontal grip, cut through the ply, was further strengthened by a 1.8 cm-thick piece of wood glued behind it on the reverse of the shield. Iron ring loops on the inside of the shield, fastened on outside, were interpreted as being for a hanging/carrying strap.

Although we have evidence from literary sources and visual imagery for the use of shields (in their changing shapes and methods of use), the archaeological evidence of actual shield remains, as discussed, is practically nonexistent, with the notable exception of the Bomarzo shield (for the beginning of the period) and the Fayum shield (from the later Republican period). While there is a wide temporal gap between these two examples, the same can also be said also for quite some time later, with the next significant remains found being those from the early Imperial period, in the form of the Doncaster shield (although heavily combusted), the Masada fragments (if indeed these are correctly identified as 'shields'), and the later, wonderfully preserved remains from Dura Europos (all as discussed in subsequent chapters).

IV

SHIELDS OF THE EARLY IMPERIAL PERIOD

The Doncaster shield and the Masada shields, as described below, can all be dated to the earlier Imperial period. Although none of these can be identified as being the standard, typical 'Roman' shields as used by the legions, their full shape and decorative features being in all cases unclear, they still nevertheless provide important evidence as to constructional methods, materials used and possible decoration. The interpretation of the finds also provides for suggested cultural origins, function and status of the user, and possible use in ceremonial or combat situations (from discussions of weight, manoeuvrability and level of body protection), which will be discussed in detail below.

THE DONCASTER SHIELD (IMPERIAL PERIOD C. AD 87)

The Archaeological Evidence

In 1971, at Doncaster, England, during excavation into the defences of the Antonine-period Roman auxiliary fort of *Danum*, in advance of the construction of an inner ringroad, the burnt remains of a shield were found. Again of plywood construction, despite similarities to those shields found at Dura and the Fayum, it was interpreted and reconstructed as a flat, long, oval/rounded rectangular auxiliary cavalry shield (Figs 33–34; Buckland, 1978, 247).

Buckland suggested that the fort had seen a period of neglect under a caretaker garrison, while the main body of troops were away fighting in Scotland under Agricola. After partial withdrawal of these troops from Scotland, in or shortly after AD 87, lowland forts such as Doncaster would have been remodelled to again house full garrisons, with smaller stonewalled forts being constructed in the late AD 150s or early 160s (*ibid.*). The rectangular Antonine fort, 2.37 ha in size, enclosed by stonewall defences, represents a later period of occupation from a larger (2.6 ha) Flavian to Hadrianic fort. Evidence of widespread burning was found between two phases of the earlier fort, although this was concentrated into defined bonfire areas of orderly demolition, followed by immediate reoccupation. Coins of Vespasian and Domitian,

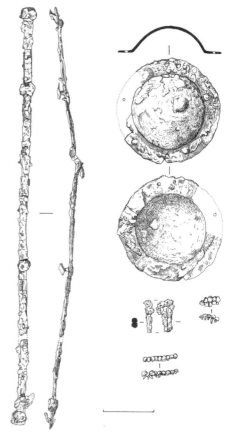

Fig 33. Doncaster finds. (Artwork by J. R. Travis, after Buckland, 1978, 252, fig. 4)

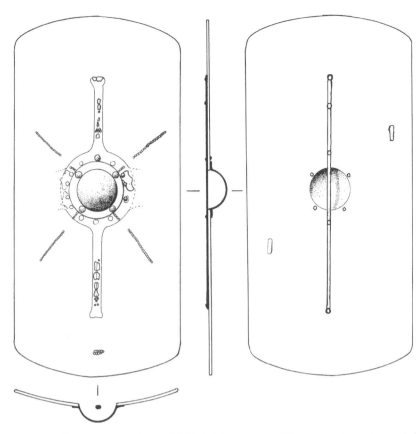

Fig 34. Illustration of Doncaster finds on shield: (a) front view; (b) reverse view. (Artwork by J. R. Travis, after Buckland, 1978, 254 & 255, figs 6 and 7)

the commonest being asses dating from AD 86 to AD 87, provided dating evidence for this phase of construction (*ibid.*).

The shield was found face down, aligned north–south on the edge of one of these clearance bonfires. The area of burnt daub and charcoal lay under a section of the later Antonine rampart, on the edge of an internal road of the early fort, partly disturbed by a construction trench of the second phase. The shield consisted of iron boss, iron handgrip, remains of brass strengthening/decorative shield furniture, and carbonised traces of up to half of the shield board, along with traces of organic surfacing (Fig. 33; Buckland, 1978, 249). Before lifting in a 30 cwt block of earth, set in plaster and encased in a metal framework, for controlled excavation off-site, gamma-ray photographs were taken of the underside surface, by boring and inserting a cobalt 60 x-ray source (*ibid.*). However, this did not reveal any clearer information of iron or bronze fittings, including any traces of any form of the brass shield edging frequently found fragmentary and as scrap on other military sites.

The Doncaster Shield Board Construction
There had been substantial heat distortion, springing apart the plywood layers during the burning process, resulting in displacement of the boss and handgrip, and loss of clear definition of the edges of the board. The boss and handgrip were no longer attached to the shield board (the grip lying at an angle of 35° to the side of the board), having sprung away during combustion, taking some of the wood with them. By consideration of the area of shield board lacking in carbonised material, Buckland felt able to approximate the position and direction of the grip and boss (0.6 m from the southern end and 0.65 m from the 'top', northern end). In his view, there was 'no doubt' that the iron handgrip had been positioned

vertically (Buckland, 1978, 248). The actual shape was impossible to define accurately, although Buckland felt able to make an estimation based on his initial visual interpretation, the decorative features of the reconstructed shield in Doncaster Museum then being the product of conjectural interpretation of multiple heavily corroded and displaced brass fragments (Fig. 33).

The shield board was estimated by Buckland as being 0.64 m wide by 1.25 m maximum length, based on the length of the handgrip and the carbonised fragmented remains, although the actual remains only consist of most of one side and just over half of the base (Buckland, 1978, 251). The side of the board is reported as straight, with possibly a slight curvature (1.1 m long). Buckland suggested a slightly rounded end shape, strongly rounded at the outer edge and following a more gentle arc at the centre, which he derived by drawing a line tangential to the apparent serrated bottom edge as excavated (Buckland, 1978, 251). However, the remains were heavily combusted and would likely have been subject to distortion and shrinkage from the burning process. The full width or length is not preserved intact, and fire shrinkage would make these sizes less accurate, so sizes can only be estimated. The remains were also compressed by the weight of overlying soil deposits, so this 'flat' shape cannot be proven conclusively.

Traces of wood adhering to the reverse of the flange indicates a grain direction perpendicular to Buckland's interpreted 'top' edge, suggesting planking following a horizontal direction on the outer front surface (planking reported as 103 mm wide by 3.8 mm thick). Similarly, that on the handgrip suggested to Buckland horizontal banding on the inside surface (3.2 mm thick). He suggested the use of alder, a relatively soft wood, for the horizontal planking, with a harder wood, such as oak, for the inner vertical layer, estimating its construction from up to six planks of approximately 110 mm width (Buckland, 1978, 251).

The shield had been extensively burned, in some areas surviving only as a silhouette. Initially, a plan was made of the surface area, all visible charred remains and metal fittings, and following minimum consolidation, the entire soil block was lifted and removed for examination in the Ancient Monuments Laboratory in London by Leo Biek (Biek, 1978).

X-ray examination was used to locate the board edges and any unseen fittings, which were then removed, along with sampled organic residues (charred wood, 'bubbly char' and traces of possible 'leather') for further detailed examination. Areas of 'organic' material were observed to be preserved on the metal fittings (boss, grip, iron and copper studs) on those surfaces that had been in direct contact with the shield board. On visual examination of these remains, Biek interpreted that the shield board had been constructed of three-ply layers of wood, glued together (the 'bubbly char' being remains of the glue), and covered in leather, from possible organic traces under stud heads (Biek, 1978, 268).

At least two well-defined wood directions were seen, at right angles, with traces of charred wood on the handgrip and the back of the boss being horizontally grained in both cases. According to Biek, botanical identification of the charred wood was 'difficult', although with the uses of a scanning electron microscope he felt able to identify two principal kinds of wood: vertical grains of oak (*Quercus Sp*) and horizontal grains of alder (*Alnus Sp*). A few small pieces of softwood (*Pinus*) were considered to be part of the general bonfire and not part of the shield construction (Biek, 1978, 269). Using the interpretation of the handgrip as vertical, wood traces on its reverse at right angles suggests that the inner face was formed of horizontal strips. As traces of wood on the back of the boss were also horizontal, Biek suggested that the shield board had been of three-ply construction, with the inner layer vertical (Biek, 1978, 268). However, if the handgrip were not to be interpreted as vertical, the construction may have been only of two-ply wooden layers.

On one side of some of the charred wood could be seen traces of a black, glossy, 'bubbly char' of blisters in a smooth matrix, formed by escaping vapours when molten. To test his interpretation of this 'micro-cellular structure having formed as a result of heat effects on glue', Biek hoped to use infrared analysis, thin-layer chromatography and amino-acid analyser. The presence of the amino acid hydroxyproline would have been indicative of protein/polypeptide residue of animal origin. However, the infrared analysis found only slight evidence for polypeptide degradation, and although the thin-layer chromatography and amino-acid analyser confirmed the presence of amino acids in low amounts there was no evidence of hydroxyproline. It was not therefore possible to rule out bacterial or vegetable contamination from the ground. Similarly, it was not possible to prove conclusively that residues, which resembled charred leather, were not animal glue, due to their common animal-based origin (Biek, 1978, 269).

The handgrip on the Doncaster shield was formed from a D-sectioned iron bar, 0.8 m long, 17 mm wide by 10 mm thick, flattened at both ends to 24 mm by 22 mm. It was fixed to the board by six square-shanked (5 mm) rivets with sub-circular heads (21 mm). These passed through holes in the handgrip bar (which were narrow on the outer edge and widely splayed on the flat inner surface), and hammered out onto the curved surface. The rivets were spaced at intervals ranging from 115 mm to 200 mm, with the largest gap across the boss to allow space for the grip. Microscopic traces of possible 'organic' materials were found adhering to this central part of the bar, which Buckland interpreted as leather handgrip binding (Buckland, 1978, 249). An iron 'ferrule' (15 mm wide and 3 mm thick) was attached to one end of the bar. This was suggested by Buckland as being for holding one end of a carrying strap, the other end being fixed to an eyelet loop (65 mm long and 35 mm across the loop) found in association with the shield board (Buckland, 1978, 249). Additional traces of material found adhering to the inner surface of the bar were fragments of bronze sheeting (interpreted by Buckland as suggestive of decoration to the inner face) and traces of wood with grain aligned perpendicular to the direction of the grip. This then became the basis for the reconstructed vertical handgrip, on the assumption of a three-ply board, both outer surfaces being horizontally grained (Buckland, 1978, 257).

Doncaster Shield – Reinterpretation
On examination of the report of the excavation and subsequent reconstruction of the Doncaster shield by Buckland, a number of issues spring to mind, which could bear further discussion and possibly alternative interpretation.

The remains of the shield that were found were not complete. The shield had been partly destroyed by later intrusive activity in antiquity and was missing the top and one side edge. The boss and handgrip had sprung away from the shield board during the burning process, causing significant damage to the board structure at that time. In addition, the board had been distorted through burning and compression as a result of burial under substantial overburden of earth. Therefore its size and shape were estimated, as was also the direction of the handgrip.

The burning process would also have caused shrinkage of the wood which may have affected the length, shape and, to an extent, depth of the planks. It is unlikely that the overall thickness of the shield would be materially incorrect (estimated by Buckland as being approximately 10 mm), this being indicated by the depth of the clenched studs. However, combining his determination of the outer surface at 3.8 mm (from deposits adhering to the boss) with that from the inner surface of 3.2 mm (from the handgrip deposits), the remaining vertical inner layer could have only been around 3 mm deep, even allowing for glue and any leather/organic surface covering (Buckland, 1978, 251). If any shrinkage had taken place in the depth of these outer surface layers, it would mean even less were available for a third inner layer, and in which case, it is possible that there may have only been two layers (as on one of the fragmentary plywood examples from Dura and on several of the examples from Masada); the outer being horizontal and the inner being vertical.

This would then draw into question the direction of the handgrip, which may not have been vertical, as suggested by Buckland, but may have been horizontal, as with both the Fayum and Dura examples. At Dura, despite the discovery of six such iron bar grips, there has been no evidence for any orientation other than horizontal. As the grip on the Doncaster example had sprung away from the board during combustion, coming to rest at an angle midway between either direction, either interpretation could have been made. Furthermore, although the shield has been interpreted as though it were flat, it is still possible to reinterpret it as being curved in shape, again as following the examples from Fayum and Dura. From personal observations by J. R. Travis of the shield in its unconserved state in 1971, while still within its block of earth, and as confirmed to him at the time by curatorial staff at Doncaster Museum, and being consistent with the published section drawings, the shield had originally been believed to be curved (J. R. Travis, pers. comm., 2000). The later reconstruction, as originally displayed in Doncaster Museum, had been made on a flat board, the remaining parts of the surface decoration laid upon this (the majority of which is now missing). Further conversations with curatorial staff had advised that this 'flat' reconstruction was chosen for ease of manufacture, and to allow for better display of the decorative fragments (J. R. Travis, pers. comm., 2000).

It is possible that the handgrip, if oriented in the horizontal direction, may have originally been curved in shape (as also must those similar examples from Dura), now being broken in several places, and having been flattened/misshaped under pressure of burial, causing it to snap at the weakest points. This would then have implications for its interpretation as being 'cavalry', the horizontal grip being more suited to infantry. Logically, the springing away from the board of the handgrip during combustion would then argue for both horizontal use and of a curved shape. If the grip had been vertical, with either shaped board, it would not have been under any particular stress. However, if aligned horizontally on a curved shield, on combustion the board would have distorted away from its unnatural, man-made curvature, tearing away from the handgrip, causing it to spring away into the position it was later found.

Buckland's interpretation was that the shield had been deliberately thrown onto the bonfire for disposal, after some of its decorative features had been removed. This he saw as additional evidence that the shield was not new. However, although not a great deal of the decoration remained, there were still many serviceable parts left on (handgrip, boss, etc.). It is difficult to accept that more effort would not have been made to remove these for reuse prior to disposal, and so it is also feasible that the shield may have possibly fallen into the fire accidentally. The 'decorative' sheet bronze pieces described as being found in association with the outer surface of the shield were in a poor state and disintegrating when first excavated. Some decorative features may also have been lost at the time when the handgrip and boss sprang away explosively during burning. Much of the excavated decoration is now no longer available (P. Robinson, pers. comm., 2004), possibly having disappeared through corrosion since excavation (the original display by Doncaster Museum featuring far more remains than at present).

Buckland reports finding wood traces on the grip and boss, suggesting that they must have been fixed directly to the wood surface. However, when he reconstructed the shield, he put both the leather covering and the decorative bronze sheet under the boss, using it to help hold them to the board, which is a more logical method of construction. It is possible that when he reported finding traces of wood under the boss flange, the leather layer may have been concealed and not been noticed. Again, on the back of the shield, it would be more logical to use the metal handgrip to help to fix down the leather covering, and this would not necessarily be evident under the trace remains of wood adhering to the grip as it would be heavily mineralised. If this were the case, it would further strengthen the case for only two layers, not three, in the board construction, and for the use of a horizontal grip, leading by default to a non-cavalry interpretation.

The non-curved shape of the back of the boss was interpreted as being evidence that the shield was flat (Buckland, 1978, 251). However, it is quite possible to fit a 'flat' boss to a curved shield, as the curvature is more likely to be greater at the outer edge than at the centre. The functional purpose of the curvature is to allow the shield to wrap around the person, with the central part, being less curved, providing the maximum area to stand behind, but with some side protection and deflection. Although the curvature of the legionary *scutum* from Dura Europos appears quite marked, even in the central part, this is now recognised as being exaggerated by overzealous conservation (James, 2004, 182). Although the Doncaster shield now presents in its flattened state, after centuries of compression from burial and distortion through burning, there is no conclusive evidence to exclude the possibility of similar curvature to other plywood shields, for example from Dura Europos and Fayum, which otherwise exhibit many similar features.

Buckland acknowledged in his reconstruction that it would not be possible to completely reproduce the full range of materials available to the Roman craftsman, but attempted to find the closest equivalent. He chose to form the boss from a cold swathed sheet of mild steel, as he considered that the Roman-period metal would not be of sufficiently uniform texture for cold working, splitting along incomplete annealing, and would have had to be forged hot (Buckland, 1978, 257). It may be easier to commence the shaping process while hot (for example, forming into a soft bed provided by a sandbag covered by leather to reduce splitting and distortion), although it is quite possible, and even may be preferable, for finishing to be done cold in order to more safely hold the metal. The argument for hot working may have some validity if the available material quality were poor. However, recent analysis of Roman armour by Sim, including that from Carlisle, found its quality to be surprisingly pure, with fewer impurities than thought previously possible (Sim, pers. comm., October 2005), so it may have been possible to cold work the whole process from scratch.

Buckland compared the handgrip to similar examples from Newstead (Fig. 3; Buckland, 1978, 257; Curle, 1911, fig. 8), describing the shape of the rivet holes, widely splayed on the inside of the grip, as 'enigmatic'. However, it would not be anticipated that the rivet holes could be any other shape. A uniform hole would only be possible if the metal was drilled, and it would not be logical to attempt to drill metal of this thickness (10 mm). It is far easier, even with relatively thin metals, to punch through when modern electric drills are not available. With thick metals the only reasonable way to penetrate would be by punching through while hot. This would inevitably then make a hole wider at the side of origin, to accommodate the width of the punch/swathe, with a smaller exit hole. The larger hole would then be concealed against the board, allowing the rivet to pass through and be securely clenched at the smaller exit.

Buckland (1978, 257) acknowledged that soft iron rivets, rather than the harder mild steel in the replica, would have minimised damage to the board on construction. However, there is also an additional bonus from the use of softer materials in their easier removal in event of inevitable repair and maintenance.

He recognised that his compromise on the use of modern plywood for the shield board would produce a shield of different strength, weight and handling properties to the oak and alder original. The hard oak would have given strength, with greater flexibility and impact absorption from the softer alder, the modern ply in comparison being more easily damaged during construction. He describes the thickness of the original planking at 3 mm as 'optimal', although he fails to consider the level of expertise demonstrated in the Roman ability to produce this planking of such uniform thinness, in lengths of over 1 metre. By choosing to interpret the shield as 'flat', Buckland (1978, 257) avoided any additional problems of how otherwise to achieve the necessary curvature.

To have produced a curved shield board from flat, modern ply would have involved use of either a press or of bending the wood over a frame, possibly after steaming to soften (although this could have caused the glue holding the laminae to soften and fail). The board would then have required the addition of cross members to prevent the board reverting to its original flat shape. However, the original oak and alder ply could have been constructed within, or over, a pre-shaped frame, in which case the layers would have no natural tendency to revert to flat and the board would retain its curved shape.

Although Buckland interpreted traces of 'leather' on some areas of the shield board, Biek (1978, 269), in his analysis of this 'bubbly char' material, had been unable to differentiate between leather and animal-based glue, each being of common origin. Indeed, even testing for protein/polypeptide residue had been inconclusive in confirming 'animal origin' at all, so this interpretation of leather covering is somewhat tenuous, being based on subjective visual interpretation, rather than any scientific analysis. Leather shield covers are known (for example, that from Vindonissa), although these are removable protective covers and not an integral part of the shield's construction. However, similar shields, from Dura Europos and the Fayum, do show evidence of a layer of flexible organic material glued to a plywood base. In the case of the Fayum shield this has been identified as a woollen material, woven or felt, although clearly little remained. This flexible outer layer, whether of leather, or wool covered by a woven material (strengthened, hardened, waterproofed by a layer of gesso and paint), would have increased the capacity of the structure to absorb impacts. It would also have served to protect the shield from the elements, and provided a surface for identifying decoration, either painted as at Dura, or using applied metal sheet and studs as at Doncaster.

In his reconstruction, Buckland chose to use leather to cover both the front and rear surfaces of the shield, to provide this flexible, decorative layer, indicating that a single modern cowhide should provide sufficient material, approximately 2 square metres (Buckland, 1978, 259). However, he does not describe what type or thickness of leather was used, or what curing method was employed. There is a great variety of potential leather 'types', with a range of properties that could have been exploited, from simple absorption of a blow through to actual physical strength. Boiled leather and rawhide, for example, can provide a material almost as hard as metal. However, those shields found to have been leather-covered at Dura are described as using thin skin or parchment, as were those found at Masada, so perhaps kid or sheepskin may have been more appropriate.

Buckland reported difficulty in stretching and gluing leather over the board; wetting to mould it to shape and making use of decorative shield furniture (boss, studs, etc.), to hold it in place. He used a 100 mm margin turned over from the front to sew to the backing sheet, 'after the fashion of the shield covers of Vindonissa' (although these were loose, removable covers), but found it necessary to cut darts from the margin at the curved ends, to reduce bulk. He recognised that this shield, unlike others from first- and second-century AD European Roman contexts, had no evidence for more durable bronze edging (as that in Fig. 5), although he considered that this may have been a tactical choice for weight reduction, or signifying a more ritual use.

However, even if the shield was intended to be ceremonial or ritual, and was never meant to be used in anger, it would still need more substantial edging, if only to protect from everyday scuffs and knocks, as seen on the highly decorative examples from Dura, which James interprets as being more probably for parade or cavalry sports use (James, 2004, 166). In comparison, on the Fayum shield, a series of holes around the edge could have been used for stitching on an extra strengthening strip, and not just to fix the front and back surfaces, as was also found to have been done on the Dura examples (not only on the plywood shields but also on those which were plank-built). It is possible that, had the Doncaster shield board not been so efficiently burnt, we may have found traces of similar edging holes. With the front and back surface covers cut to size, or with minimal overlap, an edging strip of leather could have been stretched around and stitched through, helping to anchor the front and back covering. This could then be of a thicker hide that than used for the main body. If boiled before application, it would be soft enough to stretch into place and would then harden as it cooled and dried to provide an edging sufficiently hard to withstand most knocks. The numerous studs found scattered across the board, in places still in association, in groups, as with the 'lunate' group, or the line of nine studs radiating out from the centre, would have served as anchorage for the outer covering, in addition to any glue used. If these had then been used in combination with flat metallic sheet motifs, the flat sheet would have provided anchorage against which the studs could be peened, while spreading the load of stress forces tending to pull the rivets/studs through to the other side.

There is insufficient apparent design remaining to be seen in the location of the many studs, due to their being dislodged during burning, to be able to offer any accurate reconstruction, and it would be unfair to criticise that by Buckland, as it is purely representational and much as good as any other. However, numerous examples can be seen of Roman shield designs, from monumental sculptures, such as the Arch of Orange and Trajan's Column (although of a later date), and some decorative motifs do appear on these with relative regularity, particularly the eagle wings and lightning flashes, and it is possible therefore that the radiating linear feature could be reconstructed as part of one of these lightning flashes. Many other smaller motifs are occasionally seen, which may have been painted on, or which may have been applied bronze pieces. The small fragments found here on the Doncaster shield could also be from similar applied motifs. There is no evidence to support the proposal that all of the shield may have been covered by bronze, as it would be expected that more would then have remained. If that had been the case, the argument for the leather covering then weakens, as there would have been little need to use leather at all and the bronze would probably have been extended to cover the rim.

Buckland suggested that the boss on the reconstructed shield should be positioned above centre. This then caused the lower edge to tip in towards the legs of the user, resting at an 'angle of about 30° to the horizontal', which he finds 'optimal' and describes as 'well balanced' (Buckland, 1978, 259). He draws analogies to the shields carried by early Iron Age (late fifth-century BC) warriors on the northern Italian Arnoaldi *Situla* (Buckland, 1978, 259; Lucke, 1962). However, this seems rather to be an attempt to justify his asymmetrical interpretation of the evidence. A shield that tips in towards the user's ankle could hardly be described as 'well balanced' and would be more likely a hindrance to its user, causing him to be constantly in danger of tripping.

In his discussion of handgrip direction, Buckland recognises that a vertical handgrip may be more beneficial to cavalry use, although his reconstruction, at approximately 9 kg in weight, would be rather large and unwieldy for use on horseback (Buckland, 1978, 259). This is particularly significant in view of his claim that his reconstruction would be lighter than the original due to the use of modern plywood (Buckland, 1978, 257). The legionary *scutum*, however, with its horizontal handgrip, is better suited to the closed-rank formation fighting usually adopted by the legion. The horizontal grip allows the user to raise shields more easily to form the *testudo* and to lock shields to form a shield wall (methods still in use by modern riot police). It also allows the user to 'punch' to the front, serving an offensive rather than defensive role, using the shield and its boss as an additional weapon. However, although the horizontal grip is beneficial for closed-rank fighting, the vertical grip may be more suitable for lightly equipped front-line skirmish troops – a role more frequently assigned to the auxiliary, but which may equally have been met by lightly armoured legionary troops.

There is therefore no conclusive evidence to support an auxiliary cavalry interpretation of this shield, a legionary or auxiliary infantry use being equally possible. Nor is there any conclusive evidence for the shape or decorative appearance of the shield, the remains of the board being highly distorted from the combustion process, the majority of the decoration being lost prior to deposition, and much of the remainder having been lost post-excavation, any interpretation now being entirely speculative. However, comparison with the remaining similar plywood shields from Fayum and Dura would rather support a curved shape than flat, as all indications suggest stronger similarities to curved Roman plywood shields than to flat, plank-built Celtic shields. Its size, at 125 cm high by 64 cm wide, is very similar to that of the Fayum shield (128 cm by 63.5 cm). Further, the curvature of the top and bottom edges (according to Buckland's interpretation) and also the method of their basic construction is comparable, both shields using similar plywood construction and a central reinforcing *spina* on the front, so were it not for the interpretation as flat rather than curved, both shield boards would be almost identical. The only notable difference between the two shields is in the execution of the boss (*umbo*), and mindful of the difference in dating contexts between the two, this is probably indicative of a developmental progression (Fig. 1). The Doncaster shield (with its round metal boss, common in Gaul from the first century BC) then perhaps falls into a transitional stage between the traditional curved Republican *scutum* (as represented by the Fayum shield) and the later, slightly more rectangular, semi-cylindrical shield (as represented by that from Dura Europos).

On further consideration of the Doncaster shield reconstruction by Buckland (1978, 254 & 255, figs 6 & 7; Fig. 33), the interpretation of the small, moon-shaped alignment of what was thought to be decorative studs on the front surface of the shield board was also noted with interest. However, as a result of the study of the later shield board material from

Illerup Ådal (Ilkjaer, 2002, 99; Fig. 42), an alternative possibility springs to mind. On several of these plank-built shield boards, small, lens-shaped attachments can be seen, edged and fixed to the board by similar studs. These do not appear to have been decorative, as their positioning seems to be quite random. One interpretation that may fit these features would be for patches covering areas of combat damage, the 'lens' shape approximating to the shape of the hole caused by a sword or projectile point. It is therefore possible that this similarly shaped collection of studs may also be residual from a combat repair. If correct, this would further strengthen the interpretation of the shield as having been old, having seen substantial use, rather than its being a valuable 'war trophy'.

THE MASADA SHIELDS (IMPERIAL PERIOD C. AD 73)

Masada – Background to the Evidence

Masada, set on top of a flat-topped, rocky outcrop on the eastern edge of the Judean plateau, forms a natural fortress, its surrounding sheer faces approachable (barely) on two sides: from the narrow, winding 'snake path' to the west, and from the west along a spur of land to part way, then along dangerous 'goat paths' (Pearlman, 1966, 10). These natural defences were then augmented by Herod the Great, who established a range of palaces on the top in the first century BC, by addition of a substantial encircling 'casemate' wall, along with water cisterns and storerooms, where he stockpiled food, armour and weapons, in order to be able to withstand potential siege.

Masada has become an important iconic symbol of bravery and patriotism to the people of Israel (to the extent of a tradition for army recruits swearing oaths of allegiance in ceremonies held there), derived from Josephus' reports of the mass suicide in AD 72 by *Sicarii* rebels and their families, besieged by Roman forces, who chose death rather than enslavement. However, the reports of Josephus are not necessarily totally accurate. As a previously Jewish military commander who had transferred his allegiance to the Romans, his recollections would not be without bias, their accuracy further compounded by his not actually being present during the events depicted, his information being obtained second-hand from Roman sources. The lengthy recantations of speeches by participants, the *Sicarii* leader Eleazar in particular, would have been at best based on recollections of survivors (of whom, by his reports, there were few, and being women and children, would have been unlikely to have heard them), or were narrative inventions of artistic licence intended to set the scene (*Bell. Iud.*, 7.329–405).

With the hope to confirm the accuracy of these legendary events, archaeological excavations were carried out on the site in 1965 by the Israeli archaeologist Yigael Yadin, and although these findings have been partially published since that time, the final report, including details of military equipment found, is only more recently made available (Stiebel & Magness, 2007). However, because of the obvious political importance of the site, the potential for bias within the reports is vast and this must be considered when assessing any interpretation of the finds.

The arid conditions at Masada helped to preserve a range of materials, including leather, cloth and wood. The military equipment found mostly consists of arrowheads, as would be anticipated in a siege situation, but also includes elements of scale and laminated armour, and fragments of shields, although some of these had been interpreted in earlier reports as 'leather shirts', 'baskets' and 'parts of a ceiling', so their current attribution as shields may be controversial (Fig. 35; Stiebel & Magness, 2007, 16–22, Plates 12–22). If, however, the interpretation of these artefacts as shields is accurate, their apparent structure would have implications for methods of shield construction, decoration, and also for the way that they were used.

Yadin (1966) and Pearlman (1966) both describe the rebels at Masada as Zealots, and it is possible that the first rebels who fled there to take arms from Herod's armoury may indeed have been Zealots, as confirmed by Josephus (*Bell. Iud.*, 2.433–434). However, the second

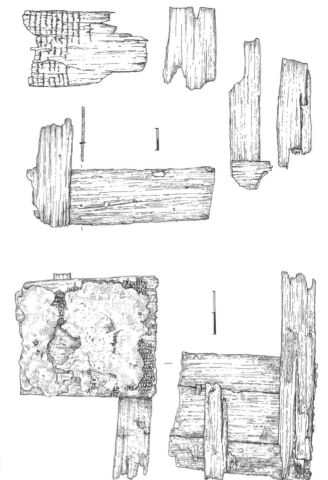

Fig 35. Plywood shield board fragments from Masada, showing chatter-pattern tool marks on wood and small pins for holding planks in place. Also showing surface coating of coarse-weave fabric and leather. (Artwork by J. R. Travis, from Stiebel & Magness, 2007, plate 19)

influx of rebels under Eleazar appears to have been of a different origin, being described by Josephus as *Sicarii* (*Bell. Iud.*, 7.295–299). There is a fundamental difference between these two types of rebels (and who were known from reported events at Jerusalem to not happily co-exist, making the simultaneous presence of both groups unlikely), in that the former, although more militant than the mainstream inhabitants of Jerusalem, were not as fanatical as the latter, who were considered more as brigands and assassins. It is unlikely, therefore, despite Josephus' account of Eleazar's choice of death before slavery (*Bell. Iud.*, 7.329), that slavery would have been an option on offer. This then does leave open the question of whether the final events were entirely as he depicted, and as to whether the heroic mass suicide actually happened, or whether the rebels fought to a man (as would be more in character), and were defeated in the final assault, their arms and armour being more suited to siege defence and distanced assault (archery equipment and scale cuirasses, with little to no evidence of helmets, apart from a few questionable 'helmet handles'), rather than for close-quarter, hand-to-hand combat against heavily armed Roman legionary troops.

Dating
The other area of concern with material from the site of Masada lies with its dating context, in the possibility of its attribution to the time of the rebel siege, to stockpiled Herodian arms, or to earlier or later Roman garrisons.

Herod, as a ruler, was not popular among his own people. His family were not native to the region, being Edomites, from the neighbouring southern region of Idumea. When the region was conquered by Judea in the second century BC, Herod's grandfather had been appointed as its governor, being later succeeded by his son, Herod's father, who then retained his position as a Roman 'client' under Pompey in 63 BC, Herod himself becoming governor of Galilee (Pearlman, 1966, 10). On his father's death in 43 BC, Herod had designs on the throne, but the people chose the more popular Antigonus (from the Maccabee royal family), who, in 40 BC, with Parthian help, drove the Romans from Jerusalem, prompting Herod to flee to his fortress palace at Masada with his brother and his followers. After appealing to the Roman senate for assistance, he later retook Jerusalem with the aid of two Roman legions and was established, in his turn, as a 'client' king in 37 BC. His hold on power was still somewhat tenuous (being unpopular with his people and his lands coveted by his Egyptian neighbour, Cleopatra), so he retained his palaces at Masada, building up its defensive structures, food supplies and stockpiling stores of armour and weaponry.

After Herod's death in 4 BC, the site remained under the control of a Roman garrison until taken by a group of Zealot dissidents. Josephus, in his *Jewish Wars*, reports that their leader, Menahem, then made use of some of these stockpiled arms to re-equip his men for an attack on Jerusalem (*Bell. Iud.*, 2.433–434; Stiebel & Magness, 2007, 31).

A five-year rebellion against Rome, known as the Great Jewish War, was eventually subdued by the Roman general (and later emperor) Vespasian, with the aid of his son Titus, culminating in the retaking of Jerusalem, the destruction of the Jewish temple and the deaths or enslavement of thousands of its occupants. Some of the occupants escaped, however, fleeing eastwards to the fortress of Masada, under the leadership of Eleazar ben Ya'ir and his *Sicarii* fighters, with the intention of continuing the war by way of small-scale actions of attrition (Pearlman, 1966, 10–11). Again, Josephus reports that he was able to find sufficient 'arms' and raw materials ('unwrought iron, brass and lead') to be able to equip 10,000 men (*Bell. Iud.*, 7.295–299; Stiebel & Magness, 2007, 31).

By AD 72, with the main rebellion under control, the new Roman procurator Flavius Silva turned his attention to mopping up outlying problems, moving on the Masada rebels with the tenth legion and associated auxiliary troops (10,000 men in total) and with the aid of around 10,000–15,000 'bearers' (Pearlman, 1966, 14). He set up eight camps around the base of the rock, surrounded the site by over 2 miles of siege wall and commenced building a ramp on the western side (the 'White Cliff'), topped by a 'pier' to act as a base for siege towers (Pearlman, 1966, 14). Josephus (*Bell. Iud.*, 7.405) then reports that in AD 73, after many months of siege, with the ramp completed and a Roman overwhelming assault imminent, Eleazar and his zealot followers chose to end their own lives rather than permit themselves to be captured and enslaved. Each man was to despatch his family, then a small group of leaders drew lots to kill the men, then each other, with the last man alive tasked to set fire to the site.

A variety of items of military equipment were found during the course of Yadin's excavations, possibly deriving from contexts associated with the final conflagrations (sealed under layers of burnt deposits), or from rebel material collected and piled up in cleaning operations by the later Roman garrison, including a considerable number of scales (many possibly from a single cuirass, or *lorica squamata*), and remains identified by Stiebel as 'shield fragments', although which had been previously identified as parts of a collapsed ceiling or woven baskets (Stiebel & Magness, 2007, 1–94).

The Masada Shields

The assemblage of artefacts described by Stiebel (Stiebel & Magness, 2007, 1–94) in his report on the finds of military equipment appear to derive from contexts associated with the final conflagrations (sealed under layers of burnt deposits, as with a substantial number of scale pieces, found along with the remains of three bodies), or from rebel material collected and piled up in cleaning operations by the later Roman garrison. However, the very nature of these contexts has led to potential multiple interpretations of some finds, whereby some of Stiebel's 'shield fragments' had been previously identified as parts of a collapsed ceiling or woven baskets.

Assuming Stiebel's interpretation of the finds as shields to be correct, their fragmentary nature permits an ideal opportunity to view the internal construction of the shield boards. The remains of up to ten shields were identified, consisting of fragments of shield board, leather surface coverings, some edge-reinforcing strips and a possible handgrip or reinforcing bar, although no bosses were found (Fig. 35; Stiebel & Magness, 2007, 16.22, plates 12–22). Stiebel believed that two types of shield were present: the plywood-built, legionary-style *scuta*, and one example of the plank-built (possibly wide, oval) type, as seen at Dura (with this perhaps being an early appearance of the style which appears commonplace in later centuries).

Although none of the shields were sufficiently complete to indicate their full size, shape or outline, they are nevertheless of great importance because of the information they provide on their methods of construction. Stiebel does not identify the types of wood used, but in some examples he suggests that the 'ply' was formed from 'pressed wood or plant fibres (apparently date palm)'. From the photographic images it would appear that some of the 'shields' were definitely true plywood construction, others appeared as a compressed mass, and one example consisted of just small wooden fragments (this being the 'shield' originally interpreted as part of a collapsed ceiling: Stiebel & Magness, 2007, 18, Plate 13.1).

What is particularly noticeable with all of the examples listed is their overall thickness (between 5 and 6 mm, including a surface covering of textile and/or thin, leather parchment of around 1 mm thickness: see Table 2), which is far less than had been suggested for other shields found elsewhere (such as the Doncaster, Fayum and Dura examples). Furthermore, in the majority of cases, only two layers of material (whether wood strips or plant fibres) appear to have been used, the layers laid at 90° to each other (the outer layer laid vertically), held together using wooden pegs or small copper alloy pins. This was then further consolidated and strengthened on both front and rear surfaces by the application of a flexible outer coating of a glue-soaked matrix in the form of either a woven textile (still recognisably khaki-coloured on shield no. 6, a particularly well-preserved fragment found in a cave on the south cliff), or layers of vegetable fibres, perhaps such as date palm, as seen on fragments inside the U-binding which makes up shields 9 and 10. This matrix then served as an impact-absorbing bond between the shield's hard, ply core and a flexible, thin animal skin or parchment outer layer, which was then sealed from the elements by a layer of painted decoration, still visible as traces of crimson colouration on some fragments (Stiebel & Magness, 2007, 20).

The remains described by Stiebel as shield no 7 are in fact multiple fragments of leather, found in the area known as 'the tannery' (consisting of nine very small scraps and two larger pieces), originally identified as part of a leather shirt, but which he interpreted as the surface covering that had been removed from a shield (Stiebel & Magness, 2007, 20). This may have been removed as part of a repair process (perhaps in an area used as a workshop) or, as alluded by Stiebel, for consumption, citing by example Josephus' accounts of inhabitants of Jerusalem stripping and eating leather from shields during famine (*Bell. Iud.*, 6.196–197). However, this latter interpretation would be somewhat unlikely, particularly if Josephus' reports of Eleazar's final speech are accurate, that the besieged rebels left their remaining food stores to prove that they chose death rather than slavery, not because of lack of supplies (*Bell. Iud.*, 6.329).

Although the resulting shield board thus produced was surprisingly flimsy, it was further strengthened by the application of a rigid, U-shaped copper alloy edge binding (attached using flat-headed copper alloy nails set at varying distances apart), along with a grid of reinforcing bars on the rear. Stiebel suggests that these bars fall into two distinct types: those set around the circumference of the board and one set across the mid-section of the shield which would also serve as a handgrip support. These central bars he equates to those with curved, T-shaped terminals seen depicted on the rear-face vertical axis of shields on Trajan's Column (scene LXXII), proposing this interpretation for the bar found at Masada, which he identifies as shield no. 10 (Stiebel & Magness, 2007, Plate 22:3). However, the bar being incomplete, without its terminal sections, it would not be possible to prove this interpretation, or to determine its alignment as either a vertical or horizontal handgrip, which would have significant bearing on the interpretation of the shield as either legionary-style *scuta* (which would traditionally use a horizontal grip), or perhaps an auxiliary or cavalry-style oval shield which may have used either horizontal or vertical grips respectively.

Overall therefore, the 'shields' from Masada (if indeed they are all shields) are too fragmentary to determine positively their shape, and by consequence to hazard more than a passing estimate to their use. Their basic construction is not dissimilar with that known for Roman shields, but positive identification as Roman is not possible, with a strong possibility of their belonging to the Zealot or *Sicarii* opposition, either from the Herodian stockpiles or new issues. The remains appear to come from contexts dated to the revolt, but could have belonged to either side. Their fragmentary nature, along with the lack of shield bosses and other shield furniture suggests that these are the remains of broken scrap collected together in the final clearing up operations, with any valuable, reusable metal fittings removed first. Had these been present, or had the remains been a little more complete, it might have been possible to perhaps estimate the identity of the possible owners, or at least to recognise the shield shape, whether auxiliary or legionary, whether using horizontal infantry-style handgrips or using vertical grips better suited to cavalry and light skirmish troops. However, as the Herodian, Zealot or *Sicarii* equipment would have probably been similar to that of Roman auxiliary troops, 10,000 of which were known to have accompanied the procurator Flavius Silva and the tenth legion on this campaign, identification would possibly have been only between the two basic shapes, oval and semi-rectangular.

The value of these remains therefore lies in their contribution to our knowledge of construction methods, in the identification of the small wooden pegs and copper alloy pins used to secure the layers of ply and to hold textile and leather surface coatings in place. Furthermore, the remains provide additional evidence for the use of leather coatings attached to the board itself, as an integral component, rather than as removable covers, as seen from sites in Europe. However, the most important aspect of these remains is the implication for their weight, in that they are perhaps only half the thickness that had been previously thought for Roman shields, suggesting that they would have been easier to carry on the march, and lighter, more manoeuvrable to use in combat situations.

Shield	Description	Plate no	Layers	Covering	Maximum thickness
1	Plywood	12:1-2	3	Leather	5mm
2	Plywood	13:1	2	not known	6mm
3	Plywood (with cu U binding)	13:2	2	Leather	7mm
4	Plywood	19-20	2	Textile + leather	7mm
5	Plywood	21	2	–	
6	Plank built (5-6mm planks)	15	1	Leather	7-8mm
7	Leather cover only	16-18	-	Leather	-
8	U binding (cu alloy)	22:1	2	Vegetable fibre + leather	4mm
9	U binding (cu alloy)	22:2	2	Vegetable fibre + leather	4mm
10	Reinforcing bar (486mm x 16-18mm x2mm)	22:3	-	–	–

Table 2. Shield fragments from Masada, from Stiebel & Magness, 2007, 16-22.

V

SHIELDS OF THE LATER EMPIRE

The site of Dura Europos has produced evidence of not only the mid- and later Roman Empire (with remains of the later-period round, convex shields, as also seen at Illerup Ådal), but also serves to cover the transitional period from the earlier Empire (with remains of the earlier semi-rectangular standard 'legionary' shield, along with a few less usual shield types (such as archer's pickets and large ceremonial shields). It is particularly useful as it provides information from a distinct, documented moment in time (that of its known attack and conquest), indicating a range of shield types in use contemporaneously.

The shields from Illerup Ådal, in eastern Jutland, Denmark, are of similar date to those from Dura Europos. However, unlike the latter they are not, strictly speaking, from a Roman context. The site is beyond the 'Roman' boundary of northern Germany, but at the time of deposition the area was under its influence, at least in weaponry and other easily transportable goods. The finds consist of multiple large-scale votive deposits from a water-logged bog/lake context. They have been interpreted as being weapons and army equipment, taken as spoils of war, sacrificed as votive thanks for victory, into what was once a lake, from around AD 200 (known as the 'Roman Iron Age'). In addition to the shields, the finds (totalling around 15,000 objects) include 'Roman' style swords, many with Roman manufacturer's stamps (Ilkjaer, 2002, 22–134). The shields found, although of similar design, shape and construction to some of those from Dura, along with other weaponry deposited appear to be a mixture of Roman and Scandinavian, decorated in more Germanic style.

Both of the above sites will therefore be discussed in order to consider shields of the mid- and later Empire.

DURA EUROPOS SHIELDS (TRANSITIONAL, MID-EMPIRE, C. AD 250s)

The Archaeological Evidence
Established in 303 BC as a Seleucid garrison at a natural bottleneck on the royal road along the south bank of the Euphrates, between Babylonia, Seleucia, Antioch and Syria, Dura Europos started as a small fortified settlement, developing into a major walled city by the mid-second century BC, James describing it as a 'mostly Mesopotamian civilian community with Greek aspect' (James, 2004, 11). Although currently known as 'Dura Europos', in antiquity the site was known either as 'Dura' or 'Europos', never both together. Isidore of Charax called it 'Dura', and described it as 'a foundation of the Macedonians, called Europos by the Greeks' (James 2004, 11).

The city had a troubled history, with the ruling power alternating many times over a 300-year period. From 113 BC the city was under Parthian control and continued as a Parthian stronghold through the first and second centuries AD (Leriche, 1996). Captured by Trajan during his Mesopotamian expedition, erecting a triumphal arch in AD 116, it was evacuated once more before his death (Teixidor, 1987, 187–8). However, by AD 165 it had been retaken and permanently occupied for Rome by Avidius Cassius, possibly after a siege (Leriche & Mahmoud, 1994, 411), becoming the largest of a series of Roman military bases and ports along the Euphrates (Gawlikowski, 1987). With the collapse of Parthian rule, and the establishment of the Sassanid Persians, it came to be a key point of forward Roman defence.

By AD 204 'Dura' had been made a *colonia*, the city being heavily modified and influenced in Roman fashion, with many new buildings and military inscriptions, and despite half a century of what James refers to as 'complex wars', culminating in the mid-third century AD in becoming the seat of the *Dux Ripae*, the military commander of legionary and auxiliary troops in the middle Euphrates (James, 2004, 11). Around AD 252/253 the city was again briefly taken by the Persians, back under Roman control by AD 254, besieged and finally overrun and destroyed around AD 255–256 (James, 2004,11).

The site of 'Dura Europos' was identified in 1920, with preliminary excavation in 1922–23, and main excavations carried out between 1928 and 1937, sponsored jointly by the French Academy and Yale University (James, 2004, 4). In addition to the military installations (wall, towers, siege works, etc.), the site produced a wealth of evidence for mid-third-century AD Roman military life, in the form of artistic representations found on the walls of the city, and documents. The site also produced a substantial collection of well-preserved military artefacts of arms and armour, metallic and organic, including shields, helmets, body armour and cavalry equipment, catalogued in the final report of 2004 (James, 2004).

The Dura Europos Shields

In the course of the excavations at Dura, four different types of shield were found: one large, horizontal, oval-shaped wooden shield (with no shield boss); two possible archery picket shields formed from leather/hide/rawhide, woven with vertical rods; three round/oval plank-built shields (convex in two planes); and four semi-cylindrical/rectangular plywood (three-ply) 'legionary' shields. Due to the quality of preservation of organic materials on site, several of these shields still retain evidence of their surface decoration, the high level of which suggests intended ceremonial rather than combat use (Figs 36–38).

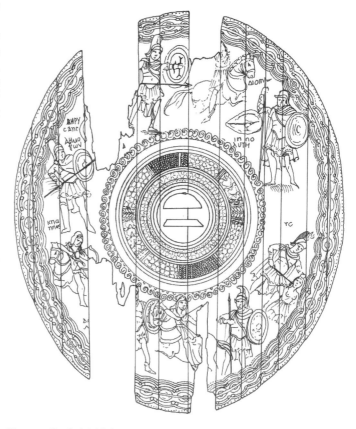

The remains of up to twenty-two shield boards were found (two of which are now missing), some with excellent examples of painted decoration, along with many other fragments also now missing (Figs 36–38; James, 2004, 159). Further, mostly in the context of the countermine under Tower 19, twenty-one shield bosses were found (although all associated shield boards were fragmentary), which then suggests evidence of a minimum of forty-three shields in total (Fig. 43). From the shield boards found, it is possible to distinguish four separate types, and to also attempt to assign boss types with some level of probability from the pattern of decoration remaining on some of the boards.

Fig 36. Oval shield from Dura Europos, shield 11 ('Amazon shield'). (Artwork by J. R. Travis, from James, 2004, xxviii, plate 7)

i) BROAD OVAL SHIELDS WITH METAL BOSSES

This appears to be the most common form in use at Dura, at least in that this is the form most represented in the archaeological record (Fig. 36). The board is a broad oval shape, averaging 1.05 m high by 0.90 m wide, and slightly convex, the centre sitting approximately 10 cm higher than the rim. It is constructed of thin, vertical planks of wood (identified as poplar, *Populus euphratica*), glued edge to edge and planed to reduce the overall thickness from 7–9 mm at the centre to 3–5 mm at the edge. James reports visible facets from axe or adze and chatter marks resulting from the use of a plane or drawknife (James, 2004, 160). The individual planks would probably have been soaked or steamed to increase flexibility, then built up over a former, the shape trimmed and the handgrip aperture cut after completion.

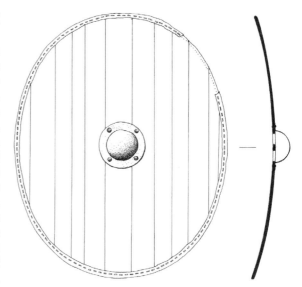

Fig 37. Oval shield from Dura Europos (shield board 619), showing plank construction, degree of curvature and applied leather edge binding. (Artwork by J. R. Travis, from James, 2004)

Two holes were cut close to the centre of the board, the top hole being semi-circular, large enough to accommodate the hand and knuckles, the lower hole being either semi-circular, to match the upper, or more usually trapezoid in shape. The narrow horizontal bridge of wood left between these two holes then formed the core for the handgrip. This bridge, being cut across the grain, would be a potential position of weakness, which was therefore strengthened by the addition of an iron reinforcing cross lath on the inside, also serving to consolidate the remainder of the vertical planks. The remains of six of these reinforcements were found, some fragmentary (similar to other examples from Doncaster and Newstead), formed from D-sectioned iron bars, with punched holes along their length which indicated that they were fixed by rivets to the board, four on each side, 100 mm to 50 mm apart. On the examples found at Dura, the central section of the bar was widened to form two 'wings' which had been bent upwards to wrap around the wooden bridge, to form a more substantial grip

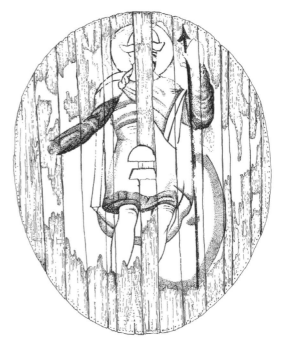

Fig 38. 'Shield of the Warrior God' from Dura Europos. (Artwork by J. R. Travis, from James, 2004, xxix, plate 9)

(James, 2004, 162). Some shields were also found to have a metal ring on the rear, left-hand, upper side possibly for a carrying strap (James, 2004, 162).

To further consolidate the shield board, and as a measure to mitigate against any potential towards vertical splitting of the planks, either along the grain or down the joints, the board was then covered with a matrix of fabric or fibre (vegetable or tendon, as used in manufacture of bows), laid at right angles to the planking, embedded within a thick layer of glue. On top of this glue/fibre matrix, a layer of either thin skin or parchment, or gesso (a plaster-like substance) provided a smooth surface for the painted decoration. Of the thirteen shields of this type found, three used skin covering over glue/fibre matrix, eight had a gesso covering over glue/fibre matrix, and two had gesso over glue/fabric (James, 2004, 162). Whether skin-faced or gesso-coated, the shields would then be edged with an additional narrow leather or rawhide strip, 20 to 30 mm wide. This was usually applied after the gesso, or overlapping the skin layer, stitched around the edge of the board using double twine threads in alternate directions, as evidenced by holes around the edge, 5 mm to 15 mm from the edge and 10 mm to 20 mm apart (James, 2004, 162).

ii) OVAL SHIELDS WITHOUT BOSS

Little is known of these two shields, as both examples are now missing. Their descriptions do not record their size, nor is there any information as to whether they were flat, curved or convex. All that now remains are the illustrations (James, 2004, 163), which suggests that both shields were oval, one was almost complete, the other only partial, with the possibility that the latter at least may have been of ply construction, covered with cloth or parchment. Both shields were decorated with a central human figure, indicating a lack of grip aperture or position for a boss on either, and also suggesting that they would have been carried with the long axis of the oval in the horizontal direction (*ibid.*). Four holes, forming a rectangle, in the centre may have served for the attachment of a grip or strap on the inside. The image of the figure on the partial shield can be seen carrying a similar oval shield, his arm passing through a loop on the back, using it as a buckler, similar to a *hoplon*-type shield. It is therefore possible that this shield may have been used in similar fashion, possibly, as suggested by James (2004, 168), deriving from the local-style desert Targe, as used by horsemen and camel riders.

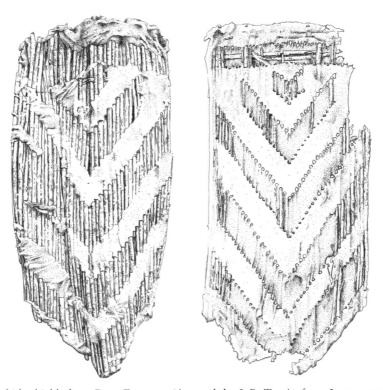

Fig 39. Rawhide shields from Dura Europos. (Artwork by J. R. Travis, from James, 2004).

iii) RAWHIDE SHIELDS, WOVEN WITH WOODEN STICKS

Three almost complete and one fragmentary examples of this shield type were found. They are formed of rectangular sheets of rawhide, pointed at one end, threaded through by approximately forty-eight straight sticks, 10 mm to 15 mm thick, to form a W or V pattern (Fig. 39). The sticks would have been inserted while the rawhide was wet and flexible, and it was then allowed to shrink and harden. The sticks are secured by the rawhide being folded over at the ends and at the sides (horizontal sticks inserted at the ends under this hem to set the shape), and stitched down. A wooden baton would have been attached at the centre, to serve as a handle. One shield measured 1.55 m high by 0.8 m wide, and the other two were slightly smaller, at 1 m high by 0.5 m wide (James, 2004, 163). James considers whether these shields may have belonged to either Roman or the attacking forces, citing similar shields found in Siberian tombs and as described by Ammianus as being used by Persians (Ammianus 24.2.10). He further considers that they would have been probably used as shelters or pickets for archers, more likely used with the point uppermost, being more easily rested on the ground on the flat edge (James, 2004, 162).

Fig 40. Semi-cylindrical plywood shield from Dura Europos (original is presently at Yale). (Artwork by J. R. Travis)

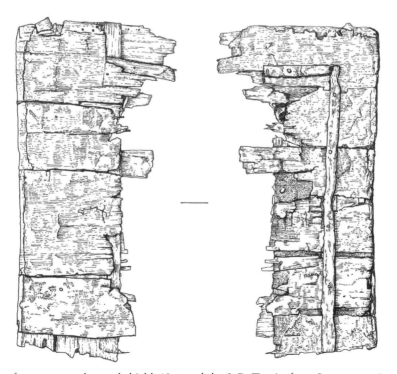

Fig 41. Dura fragmentary plywood shield. (Artwork by J. R. Travis, from James, 2004)

iv) RECTANGULAR, SEMI-CYLINDRICAL PLYWOOD SHIELDS WITH METAL BOSSES
Only three of these shields were found at Dura (one complete and two fragmentary), despite these being generally regarded as the typical Roman legionary shield type. The complete example was not found in the countermine combat context and was missing its boss prior to deposition, presumably being stored in Tower 19 awaiting repair or completion (Fig. 41). This, from its decoration, appears to have originally been intended to use a rectangular boss, fixed by rivets at the four corners, similar to the one fragmentary example found at Dura (James, 2004, 162), or the complete example as found at South Shields (Bidwell, 2001, 10).

The shield from Tower 19, although missing its boss, was otherwise complete, and particularly well preserved, uniquely even retaining its entire original painted decoration. Despite its description, the shield was not entirely rectangular, being slightly rounded at top and bottom, measuring 0.85 m wide and 1.05 m high at its maximum point. The present curvature is not representative of its original state, being a result of overenthusiastic conservation at Yale University (James, 2004, 162), the condition on discovery being fragmentary and distorted, but with a much shallower curvature following an apparent cord width of 0.66 m.

The shield is made up of three layers of thin wooden strips 30 mm to 150 mm wide, the middle layer being laid vertically (of 2 to 3 mm thickness) and the outer layers being horizontal (1 mm to 2 mm thick), of an overall thickness of 5 mm to 6 mm. The wood used has been identified as plane (*Platenus orientalis*). The two fragmentary shields also found appear to be of similar materials and construction, although one appears to be constructed of only two layers of ply (Fig. 41; James, 2004, 163).

As with the oval shields, the edges of the board were trimmed to shape after completion, and grip aperture cut as previously, with two semicircles approximately 120 mm across, forming a central bridge for the grip. On the rear of the board, a framework of four strengthening laths, 20 mm wide and 2–3 mm thick, was glued and pegged with dowels 90 mm from the edge. Across the middle, a fifth lath also served to thicken and strengthen the handgrip, being further lashed with leather or rawhide for comfort (James, 2004, 163).

Again, as with the oval shields, the board was then covered with a matrix of thick glue and thin linen fabric, before a final layer of thin, red-dyed kid or parchment. Strips of leather, 35 mm to 50 mm wide, were then sewn over the edges, overlapping the skin covering, to protect the edges from damage and also to help prevent separation of the ply, with extra strengthening pieces of leather applied to the corners (James, 2004, 163). The two fragmentary shields had again been similarly treated, both coated with a glue/fibre matrix, although then one had been skin-covered, as on the complete example; the other gesso-coated, as on some of the oval, plank-built shields.

THE ILLERUP ÅDAL SHIELDS (TRANSITIONAL, MID-EMPIRE, c. AD 200s)

Background to the Finds
The first finds from Illerup were discovered through peat cutting in the nineteenth century AD. The first excavations were carried out in the 1860s, over an eighteen-year period, at Thorsbjerg and Nydam in Schleswig, and at Vimose and Kragehul on Funen, by Conrad Engelhardt, and published 1863–69. Further excavations then took place in 1950–56 and 1975–85 (Ilkjaer, 2002, 15). Heaps of finds were then lifted by the 'preparation method', where the sides of a wooden box was dug down around the heap, then its base pushed under and the whole box of finds and earth lifted for examination at Moesgard Prehistoric Museum (where the box was turned over and excavated from other side; Ilkjaer, 2002, 38).

After the Ice Age, the lake at Illerup was 25 m deep. By 1,800 years ago this had reduced to 3 m deep, as material washed off the surrounding slopes (moraine clay), making the lake alkaline. This process then continued until the site became an alkaline bog, as it remains today. It provides good preservation conditions for bone, antler, iron and other metals, but leather and textiles dissolve. In contrast, Thorsbjerg is an acid bog where iron, bone and antler do not survive, but clothing and leather are preserved. In both alkaline and acid bogs, wood is preserved (e.g. spear shafts, shield boards, etc.), but only the cell walls. The cell's core is replaced with water, so only retains its form while under water (Ilkjaer, 2002, 21).

Around AD 203, a large fleet of Nordic warriors (over fifty ships) sailed close to eastern Jutland, landed, battled with locals and were defeated. Spoils were taken of clothing, weapons, and equipment. Clothing was shredded, swords bent, shields smashed, leather goods chopped up, all rendered unusable, bundled up, sailed out onto the lake and cast overboard (Ilkjaer, 2002, 15). Draining work in 1950s lowered the water table and the upper peat layers began to disappear. Without excavation, artefacts from the deposit would have deteriorated. By that time some had already been destroyed or had begun to disintegrate (Ilkjaer, 2002, 25).

Geophysical surveys were conducted in the 1970s to attempt to find the full extent of the deposits. It had originally been thought that all were to be located just within throwing distance of the lake edges, but it was found that the deposits were more extensive (Ilkjaer, 2002, 30). Some of the deposits appeared to have been thrown in from the shore, although other deposits were further out into the lake, appearing to have been made in heaps from boat deposits (so from multiple deposits). Some items had been badly damaged before deposit, for example one very ornate silver shield boss, which had been torn from its shield then pierced by lance and hit by sword/blade, in order to 'kill' it/render it useless (Ilkjaer, 2002, 36). In addition, some broken pieces from different heaps were found to match, so some deposits were contemporary, suggesting that large quantities had been broken up on the shore and taken out in boatloads (Ilkjaer, 2002, 30).

Dating Evidence
Dating evidence for the deposits came from coins, dendrochronology and styles of weaponry. One deposit was found to contain 200 Roman silver coins, with the most recent providing a *terminus post quem* dating of AD 187/188. Tree-ring dendrochronology of the oak in one shield also suggested that it had been felled in AD 187, with repair done after AD 205 (as the wood would have been seasoned for some time before use and as the shields had not been new when deposited), possibly bringing forward slightly the dating of the deposit. Furthermore, stylistically the swords and scabbards correspond to those on Septimius Severus' triumphal arch (with Severus being in power from AD 197–208; Ilkjaer, 2002, 48).

No clothing was preserved in the deposits, but they are believed to be contemporary with those from Thorsbjerg, where shirt/tunic and trousers (with belt tops and build on feet) were found in the acid bog. Similarly, no helmets or traces of mail were found, although it was considered that the soil conditions would have permitted these to have been preserved if present (Ilkjaer, 2002, 51).

The deposits produced large quantities of weapons, with one heap particularly producing nearly thirty spear and lance points, many with parts of their shafts still intact (these appear always to have been of ash, and produced from split and planed heartwood). Also found among the weapons was a shaft plane, with concave underside, which would have been used for repairs. In total the site produced 748 lance heads and 661 spear heads (1,409 in total). Overall, this then equated to a ratio of three to four spear/lances to each sword (Ilkjaer, 2002, 42). The similarity in styles of the spear/lance heads also suggested that deposits are contemporary. Some blades (two) were found to have been stamped with the maker's name ('Wagnijo'), suggesting some mass production. Ilkjaer (*ibid.*) reported that the stamping of blades is not normally common in Germanic contexts, but is in Roman contexts. Among the Roman-style swords at Illerup, about half were found to show maker's marks or names (*ibid.*).

Further evidence of Roman or Roman-influenced weaponry is also provided by a baldric mount (bronze, circular, with openwork decoration of an Eagle/Jupiter in the centre and letters around it, reading 'OPTIME MAXIME CON'. The inscription is a shortened form of 'OPTIME MAXIME OMNIUM MILITANTIUM CONSERVA', translating as 'May the best and greatest [Jupiter] preserve all who fight'. This mount matches and has the same casting fault as another example from England, so being produced from the same mould (Ilkjaer, 2002, 96).

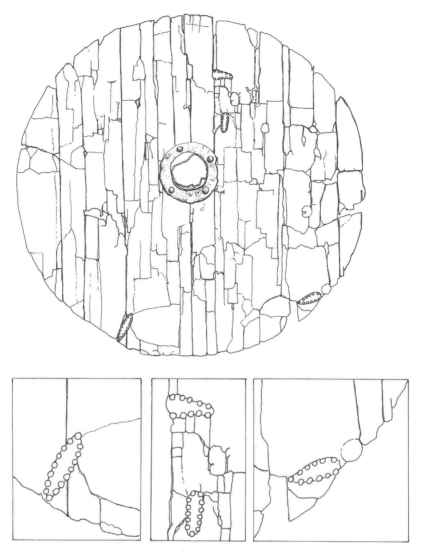

Fig 42. Late Roman-period plank-built shield from bog deposits at Illerup Ådal, Denmark; with, below, close-up of multiple lens-shaped 'repair' patches. (Artwork by J. R. Travis, from Ilkjaer, 2002, 99)

Ilkjaer (2002, 137) estimated that the quantities of equipment suggested an invading army of around 1,000 men. He further reports that around 150 belt sets were found, but as only 40 per cent of the area had been excavated, he estimated that 300–400 in total is probable, with the majority of casualties being among ordinary men with lower-status equipment (*ibid.*).

The Shields
The Illerup shields, as with many of those from Dura, were circular in shape, normally constructed from five to eight boards, usually cut from alder or oak, but also with some of poplar and lime. In the centre of each, a circular hole had been cut out for the handle. The handle was usually formed from a piece of hazel branch, covered by metal, with the hole then covered on the exterior surface by a protective metal shield boss (Fig. 42).

The boards used in the construction were very thin (10 mm in the middle, tapering to 3 mm at the rim), with the rim strengthened by addition of a thin metal binding. Ilkjaer (2002, 98) stated that he was unsure as to how the boards were held together, although he considered that the edge trim helped and that the boss would then help to hold the centre boards. However, he speculated that the remainder may have been held by organic materials, now dissolved in the alkaline conditions of the bog (*ibid.*).

In addition to the deposits being broken up into distinct 'heaps', within the deposits it was possible to identify 'layers' of different types of weaponry, suggestive of being representative of the hierarchy within the defeated army, with higher-status items at the top and ordinary-status items at the bottom. Cavalry equipment was also found, with a layer of higher-status shields, swords and baldrics.

In Illerup Site A deposits, three layers were identified. In the top layer were five high-status shields, decorated with solid silver and embossed sheet metal (boss, handgrip and rim fittings were solid silver, with decorative mounts formed as masks and conical attachments, often with gilded sheet metal). The boards were also found to have traces of red dye (cinnabar, with a chemical composition HgS; Ilkjaer, 2002, 100). In the middle level were two bronze bosses decorated with sheet metal; three iron bosses (similar to the bronze ones); and over thirty shields with bronze mounts. Parallels to these fittings were reported to have been found elsewhere in Norwegian grave contexts. In the lower level over 300 sets of iron shield mounts were also found (*ibid.*).

Some of the higher-status items were found to have been decorated with precious metals. Silver and gold had been used with 2 per cent of the finds (those that were of high status and imported). Bronze had been used with 9 per cent of finds (these appeared to be of middle status, and also to have been imported). Iron had then been used with the majority (89 per cent) of finds, this being the standard material for fittings. One shield, for example, was ornamented with twenty-two masks (of gilded silver), with a silver shield handle which had been fixed to the shield by two rivets. Rivet heads had been fashioned with an imprint of a Roman silver coin, bearing a head and the text 'DIVA FAUSTINA' (dating to Emperor Antoninus Pius, AD 138–161). The higher-status artefacts were believed to have been of Scandinavian construction, with gold, silver and filigree decoration, although in the example described above Roman imagery was used (Ilkjaer, 2002, 124).

The only artefacts of high status that were found to have been Roman were the swords (these being recognised as Roman from the maker's stamps on blades), but with Germanic-style grips (decorated with silver and gilding) replacing the plainer, Roman original, to suit more Germanic tastes. The baldrics also exhibited stylistic features reflecting their origin and status, with Germanic ones being un-adjustable, made for the individual, whereas the 'Roman' ones buttoned together or tied to rings in order to adjust for any wearer, so were probably mass produced for equipping men from an arsenal (Ilkjaer, 2002, 134).

Observations

Although not mentioned in the text, what can be observed in the photographs provided in the publication are small metal plates riveted to the boards, using small, dome-headed rivets. These do not appear to be associated with the board remains from the higher-status level 3 (Ilkjaer, 2002, 101 & 126), but can be seen on the example from level 2 (with bronze boss; Ilkjaer, 2002, 134) and on the example from level 1 (with iron boss; Ilkjaer, 2002, 99). These are randomly placed on the shield board, are mostly lens shaped (though one example is more rounded) and are more probably evidence of repair patches rather than serving a decorative function. It is not unreasonable to expect that higher-status shields would receive less damage and would also be more likely to be discarded if damage occurs, whereas the middle classes and ordinary soldiers would need to 'make do and mend'. These patches are therefore probably indicative of a shield having seen previous combat use (Fig. 42).

VI

SHIELD BOSSES

THE ARCHAEOLOGICAL EVIDENCE

The shields of the early Republican period, like the Fayum example (Fig. 32), were curved oval shapes, with a central reinforcing rib, or *spina*, and an elongated oval-shaped wooden boss, or *umbo*, covered with a moulded metal sheet (usually copper alloy), often extending beyond the *umbo* itself, a style which Feugère (2002, 78) considers may have originated with the Celts of northern Europe, or Germanic people in the fifth century BC. This style of boss then continued in use until the late Republic and early Imperial period, when a new tradition starts to appear without the central rib and the oval boss being replaced by one that is hemispherical with a round flange.

The tradition for these round *umbones* may have been borrowed from native auxiliaries, being found in pre-Roman and Roman auxiliary funerary contexts, with examples from Alesia, Slovenia and Gaul (Feugère, 2002, 78). These simple round bosses, fashioned in either copper alloy or iron, are used throughout the Imperial period, appearing on oval auxiliary shields depicted on Trajan's Column, and as found on the Doncaster shield (Figs 33–34; Buckland, 1978, 252, fig. 4), or more decorative versions from Kirkham, Lancashire (Fig. 46; Bruce-Mitford, 1964, 68, Fig. 34), Papcastle (Fig. 47) and from Halmeag, Romania (Fig. 45; Feugère, 2002, 92, fig. 108). They then continued in use with little change, being found also on the broad oval shields of the later period, with many examples found in the countermine combat context at Dura (Fig. 43; James, 2004, 168).

Some more 'exotic' versions of this basic round boss are known, some

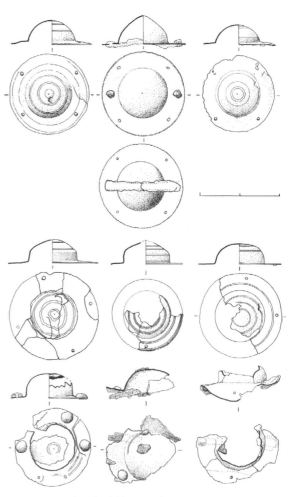

Fig 43. Circular shield bosses from Dura Europos. (Artwork by J. R. Travis, redrawn after James, 2004, 172, fig. 94)

with conical bowls, some with pointed projections at the bowl apex, and others with star-shaped projections around the flange, possibly indicating regional variations. The function of these forward projections is unknown, perhaps serving for attachment of some kind, although it is not clear what combat advantage they could provide. The projections are not

Fig 44. The South Shields shield boss ('Durham shield boss'). (Artwork by J. R. Travis, from Bidwell, 2001, 10, fig. 5A)

Fig 45. Decorated shield boss from Halmeag. (Artwork by J. R. Travis, from Feugère, 2002, fig. 108)

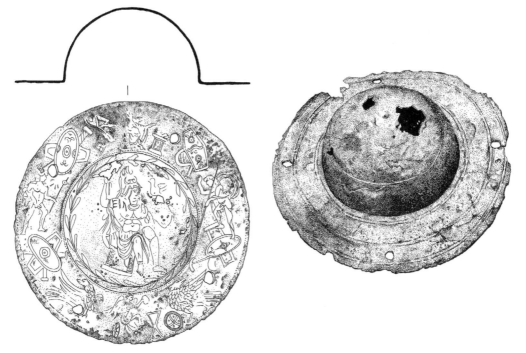

Fig 46. Decorated shield boss from Kirkham, Lancashire. (Artwork by J. R. Travis, from Bruce-Mitford, 1964, fig. 34)

Fig 47. Bronze shield boss, from Papcastle, Cumbria. (Artwork by J. R. Travis)

particularly sharp and could actually cause an opponent's weapon to become entangled rather than serve to deflect. They appear, therefore, to be a stylistic rather than functional embellishment.

At some point in the early Imperial period, the curved oval Republican shield is 'sliced off' at the top and bottom, to form the transitional stage to the later semi-cylindrical rectangular 'legionary' shield. Subsequently, this shield type often appears associated with a variant of the hemispherical boss on a rectangular flange (producing the second typical 'Roman' boss type), as seen on Trajan's Column and with examples found at Dura, and more decorative versions from Vindonissa (Feugère, 2002, 89, Fig. 101) and South Shields (Fig. 44; Bidwell, 2001, 10, Fig. 5A).

The boss on the Doncaster shield is an example of this round type, sub-hemispherical in shape, formed from 2 mm thickness iron, 200 mm in diameter overall, with a central *umbo* 144 mm diameter, rising to 28 mm high. It was fixed to the shield board through its outer 28 mm flange by four iron rivets in an X alignment (20 mm slightly domed heads and 5 mm shanks; Fig. 33). No curvature is evident in the profile of the flange, which Buckland interprets as suggestive that the shield itself was also flat. No decoration was visible on the boss due to its heavily corroded condition, and none was subsequently located by x-ray examination (Buckland, 1978, 251). Some evidence of repair was noted by Buckland, suggesting to him that the shield was not in pristine condition when lost, in that the upper edge appeared to have been re-fixed, with an additional rivet at the top centre of the flange edge.

The Newton boss is a variant of the standard round, hemispherical type, its *umbo* tapering initially at the sides, then raising to a conical cap, rounded at the apex (Fig. 48). Formed from a flat sheet of iron, slightly thinner than the Doncaster boss at 1.5 mm thick, it is however comparable in size, with an overall diameter of 194 mm and a 30 mm outer flange, where two 4 mm 'square' rivet holes can be seen (Buckland, 1978, 266). Buckland reports that triangular-shaped facets are visible on the inside of the boss, radiating out from the centre, which he considers to be caused as a result of use in the manufacture of a triangular-shaped anvil with a 46 mm lip on its broadest edge. However, it was impossible to corroborate this by physical inspection, as this boss (along with the remaining fragments of the Doncaster shield) is in an exhibition case within Doncaster Museum and access is not permitted. Although the flange is slightly distorted, perhaps caused during removal from its board, Buckland reports that the profile of the boss is curved, possibly to accommodate a curved board (Buckland, 1978, 266), which would then negate the argument for round boss types being used exclusively on flat boards.

The dating of this boss is problematic in that it was not found in a context closely related to its period of use, as with the battle contexts of those from Dura, or at the least found associated with the disposal of its shield board, as at Doncaster. The Newton boss had, by the time of its deposition, been long disassociated from its board, being reused as the lid for a cinerary urn. This urn and the two associated coffins (one lead covered, the other of stone) have been dated to the late third century AD, which indicates a latest possible dating. Buckland does not consider that the urn could have contained the original shield owner, so its true date could be considerably earlier (Buckland, 1978, 266). However, realistically it is unlikely that an old, damaged boss, with no special decorative features, would have been retained for any extended length of time once disassociated with its board. It would seem more logical to assume its reuse in the funerary context was not too remote in time from its original use, possibly being used for its original owner.

The bronze boss found at Kirkham, near Preston, in Lancashire is again of the round, hemispherical type, similar to that from Doncaster and to the majority of examples from Dura. However, the Kirkham example differs in that it features extensive incised ornament, with a central human figure (possibly of Mars) on the *umbo*, surrounded by further images of eagles, additional human figures (possibly representative of the four seasons), and crossed shield trophies (oval and octagonal and/or rectangular types; Fig. 46; Bruce-Mitford, 1964, 68, Fig. 34). A similarly highly decorated hemispherical boss is also known from a shield from Halmeag, Romania, although with a more ordered, less chaotic design, featuring a central eagle (*aquila*) surrounded by panels containing round (possibly shield) shapes, fish (possibly dolphin) and human figures (Fig. 45; Feugère, 2002, 92, Fig. 108). Although here associated with a round boss type, the eagle (*aquila*) figure could be indicative of legionary rather than auxiliary use, or may represent an exceptional citizen status in auxiliary context, as in units of *voluntariorum civium Romanorum* (Rossi, 1971, 111).

Many of the semi-cylindrical legionary shields depicted on Trajan's Column can be seen to carry rectangular bosses, and it is probable from the style of decoration that the shield board of this type from Dura would have used a similar boss (Fig. 40). The finest example found of such a boss can be seen in the Museum of the University of Newcastle. It is one of a series of objects dredged from the river bed at South Shields between 1880 and 1915, believed to have been deposited as a result of a shipwreck carrying a legionary expedition to northern Britain in the early AD 180s (Fig. 44; Bidwell, 2001, 1).

The bronze boss is decorated with panels of figures and designs pounced and incised against a Niello ground (Bruce-Mitford, 1964, 67). The apex of the *umbo* features a legionary *aquila* within a wreathlike *corona*, the rectangular flange divided into panels

Fig 48. The shield boss from Newton, near Doncaster. (Artwork by J. R. Travis, from Buckland, 1978, 265, fig. 10)

containing figures apparently depicting Mars, the four seasons, legionary standards at each side, and in the centre lower position a bull, with moon-and-star symbols above. Two panels above the legionary standards are inscribed with the owner's legion, LEG(io) VIII AVG(usta), and on the edge a less well-executed inscription identifies the owner as Iunius Dubitatus, in the century of Iulius Magnus (*c(enturia) Iuli Magni Iuni Dubitati*). The *legio VIII Augusta* was known to have been based at Strasbourg, but vexillations were recorded as being despatched to problem areas, as a form of rapid-response team. One such example is described on a second-century AD inscription to T. Pontius Sabinus, of a combined body of 3,000 men, drawn from the *VIII Augusta*, *VII Gemina* and *XXII Primigenia* being sent on an *expeditione Brittannica* under Hadrian (Bidwell, 2001, 15).

At Dura, twenty-one bosses were found, although none in association with any board. Of these, fourteen were of copper alloy, their colour varying from reddish to yellow depending on their zinc content, and only seven being of iron (Fig. 43; James, 2004, 160). This disparity may, however, be more due to differential preservation than representative of preferred use. Most bosses were circular, with roughly hemispherical bowls, large enough to accommodate a fist holding the shield grip, with only one recognisably rectangular (similar to those from South Shields and Vindonissa) although, as many are fragmentary, their shapes were not clear. The majority of the round-shaped bosses are simply shaped, with a low-profile bowl and broad flange, pierced for four attachment rivets. However, a small number of more exotic examples are known, one with a cylindrical projection at the apex (although not as extreme as in the pointed examples which predominate in Germanic regions), and two with eight-pointed star-shaped projections around the flange. Although these star-shaped bosses are not well attested elsewhere, a variant is known from Syria and a fragment of shield from

Trier is decorated with an eight-pointed design that may have accommodated a similar boss (James, 2004, 168).

As the majority of bosses found at Dura are of the round type, and only one of the rectangular type, it has been assumed that the former are always associated with the oval-shaped shield. However, whereas the rectangular bosses do appear always to be used on rectangular, semi-cylindrical shields (as seen on the *stela* of Caius Valerius Crispus of the *legio VIII Augusta* at Wiesbaden; Fig. 15; Russell-Robinson, 1975, 167, plate 469), it is not impossible that non-rectangular bosses could have been used on shields of this type. Although only one complete and two other fragmentary shields of rectangular type are found at Dura, only fragments of boards of any type are found in the combat context where the majority of bosses were located. It is therefore not impossible that more of this type of shield may have been present, its number being unrepresentative due to poorer preservation than the more substantial planks (and even these did not survive in that particular context).

The Dura bosses were attached to the front of the boards (oval or rectangular) by four rivets. One example was also found with an additional reinforcing iron grip bar attached across the inside of the boss, riveted at each side. From the positioning of this bar, it was apparent that the boss was then attached to the board with its rivets in an X arrangement, rather than in a +, similar to the arrangement in use on the rectangular boss (James, 2004, 162).

As with helmets, both iron and copper alloy shield bosses are found, although again use of iron does not commence until the end of Republic/early Imperial, with the appearance of the round, hemispherical boss. Copper alone would have been of limited use for boss production as it is a soft metal and would afford little protection for the shield user's hand, although it could have been used for the earlier shields, where sheet metal was used to cover an underlying wooden *umbo*. Therefore, the metal of choice would generally be either iron, or a much harder copper alloy, such as the reddish-brown bronze, or alloyed with zinc for the more yellow-coloured 'oricalchum', both colour variants being evidenced at Dura (James, 2004, 160). Again, choice of metal may have been influenced by strength or availability of materials, but may equally have been based on aesthetics of colour, particularly in the case of decorated examples, copper alloys lending themselves more easily to incised and pounced decoration, contrasting better with any surface silvering, tinning or applied Niello, as on the South Shields example (Bidwell, 2001, 9). As with helmet production, choice of alloy may have been dictated by the preferences and traditions of the individual craftsman, and the most readily available raw materials, and may possibly be indicative of region of manufacture.

REPRODUCTION OF THE SHIELD BOSS

The possible methods of manufacture which could have been used are similar to those for helmets: by hammering manually from sheet metal, either drawn down over a former or beaten from the inside. Casting, however, would not have been a probable method of production for shield bosses, the process being far more complicated than justifiable given the relative simplicity of the end product. Spinning on a lathe could have been a possibility, although not for the examples with the 'exotic' raised/spiked centres, but would have left some telltale evidence of a centring hole, which is not found.

The process would start with a sheet of annealed metal roughly shaped to the end product. This may have been produced by hammering from a 'bloom', working from a bar or 'pig', or a pre-imported sheet (rolled, pressed or machine hammered). There are three possible methods which could then be used to raise the bowl of the *umbo* which are broadly similar to those used for helmet production. In the first method (the simplest, requiring the minimum of equipment), the metal would be hammered from the inside, rotating in concentric circles around the bowl, starting from the centre, working outwards, beginning with a large ball-ended hammer, similar to a modern cross-pein form (Fig. 49). Once the shape has been

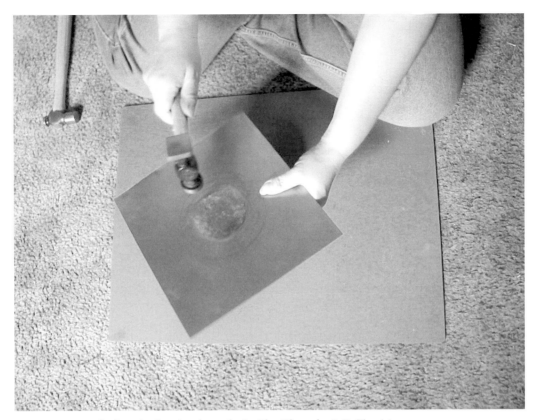

Fig 49. Raising the boss by beating from the inside. (Photo by J. R. Travis)

roughly achieved, the process is repeated using smaller and smaller hammers, to remove visible hammer marks, finishing with a small, rounded, 'planishing' hammer. The outer surface would then require surface planishing to finish off. If Buckland's interpretation of facet marks on the inside of the Newton boss are correct, this would suggest that it was produced using this method of manufacture (Buckland, 1978, 266). However, there is no evidence for triangular anvils in use during the Roman period (although lack of evidence does not necessarily mean non-existence), the marks described being equally possibly produced by the corner of a square/rectangular anvil.

The second method would be in principle as above, hammering the boss from the inside, although working the sheet into a preformed, hollowed-out shape or 'former', perhaps made of wood. The end result, as in the first method, would produce evidence of hammer marks, or 'facets', on the inside of the bowl, although the shape would be more uniform, not being produced entirely 'by eye' as in the previous method. There may also then be less need for surface planishing to even out any irregularities. James reports planishing marks evident on the inside of one example from Dura, which he interprets as being suggestive of production by sinking into a former in this way, but equally this could be indicative of either of these first two methods (James, 2004, 160).

The third method of shaping, as with helmets, consists of hammering the metal from the outer surface over a ball-headed stake. This method would be particularly appropriate for the 'exotic' examples with a raised or spiked apex. As with the Montefortino-type helmets with raised-crest knobs, the centre of the sheet is indented from the inner surface by hammering with a pointed rod. This raised point is then worked into the required shape, a process requiring a high level of craftsmanship. The annealed sheet is then hammered, again using a cross-pein type of hammer, over a ball shape of approximately similar profile to the end shape required, working around the bowl, spiralling downwards with overlapping

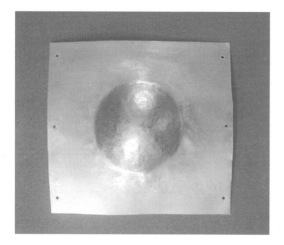

Fig 50. The completed boss with attachment holes punched. (Photo by J. R. Travis)

Fig 51. Reconstruction of the South Shields shield boss, with pounced and etched decoration. (Photo by J. R. Travis)

blows, until the approximate shape is produced. Each of these concentric hammer blows would have produced small, dished markings. By repeating the process lightly, with smaller and smaller hammers, these dishing marks are gradually 'planished' away to leave a smooth surface (Fig. 50).

Finally, the surface is ground and polished to smooth it further, and engraved or pounced decoration added (Figs 51–52). As with helmets, if the boss is being produced from iron it can be produced in one continuous process. However, if using copper alloy, the metal may become brittle and work-hardened during the hammering process, the crystalline structure within the metal aligning, creating weak points where splits may occur. This will require frequent annealing or softening, by heating to high temperatures (to randomise the crystalline structure once more), and subsequent cooling (to set the structure), to prevent splitting.

EDGE BINDINGS

It would be reasonable to expect that some form of edge protection would be necessary on plywood-construction shields, not only to help to hold together the structure of the ply and protect the shield from damage through wear and tear and water ingress, but also to increase its effectiveness when the shield edge is used as a weapon to strike an opponent. Bronze shield-edge bindings from legionary shields have been found at numerous sites (including at the Lunt fort, England, the remains of which can be seen in Fig. 5). The edging was formed from strips of flat bronze, bent into a U-shaped channel, provided with rounded protruding lugs at regular intervals, pierced for the insertion of copper rivets (the softer metal of the copper rivet allowing for easier removal for refurbishment, or removal and reuse on a replacement shield).

However, as discussed previously, remains of shield boards are less rarely found, due to their perishable nature, and none of these have been found with any edge bindings preserved *in situ*. In fact, the evidence on these extant remains suggests rather for the use of leather edge binding (which in all probability would have been of the maximum-strength leather, such as cowhide, of substantial thickness, folded over the board edge and stitched in place). It is also possible that any leather trim may have been boiled before use, to make it more flexible for application and so that it hardened on cooling (this method of hardening leather is known as *cuir bouillée*). Similar proposals have been made for the use of rawhide (as used in modern-day dog chews), which when cooled and dried solidifies to an extremely hard surface.

The Fayum shield is reported to show evidence for stitching holes around the edge which may have served to attach a wide edging trim (which, if present, would probably have been of a hard material such as leather). The shield was reported to have been covered on both sides by a thin layer of sheep felt, the exterior surface applied first, covering the shield board up to the edge; the inside surface then applied, overlapping to the front by 5 cm to 6 cm. This was then stitched down using a double linen thread, the resulting seam then folded over and glued. It is possible that a reinforcing outer layer of leather may have been added at this stage, all held in place by the same circuit of linen thread. Although, it is equally possible that a metal edging strip may have been used, there was no trace of any remains reported with the find (as there is no evidence for the shield having been to any extent dismantled before deposition, and which would have been anticipated should the shield have been complete when lost).

Similarly, the Doncaster shield was not found associated with any metal edge bindings, although remains of other surface decorations were found. However, the shield was found partly combusted in a demolition layer, and it has been suggested both that it may have been a shield at the end of its natural life, part dismantled and deliberately burned; or an active and intact shield which accidentally fell into the conflagration. The reconstruction proposed by Buckland (1978, 247; Fig. 34) has therefore been made without any edge reinforcement, leather or metal.

However, the existence of finds of bronze edge-binding strips from numerous sites (particularly in the more northern regions), along with clear depictions of metal edging on sculptural representations of both legionary and auxiliary shields, suggests that it was widely used. It may therefore represent a regional and/or cultural trend, probably originating from Celtic/Gallic influences, as bronze edge binding appears to have been standard practice on Celtic shields and British hide-shaped shields (Figs 25 & 26).

VII

DECORATION

SHIELD DESIGN CONVENTIONS

In addition to any decoration found engraved or embossed onto the shield boss (*umbo*), it is known from surviving literary sources, as previously mentioned, that the shield board itself would also have been decorated as a means to identify troop units (Ammianus, 16.12.6; Claudian, *Bel. Gild.*, 423; Vegetius, 2.18). In his discussion of legion emblems and birth signs, Dando-Collins (2010, 77) reported that every legion and auxiliary unit had its own emblem, which was used on shields as a means to identify one unit from another. He cited Tacitus (*Histories*, 3.23), who wrote that in the Battle of Cremona, two of Vespasian's soldiers used enemy shields, with the emblem of the opposing Vitellian legion, and passed through enemy ranks to sabotage their massive catapult.

In his study of Trajan's Column (now considered to be perhaps somewhat dated but not entirely without merit), Rossi suggested that this system of unit identification was then used to create a convention whereby a small number of essentially identical soldiers, but with differently decorated shields and standards, each represented a larger body of men (Rossi, 1971, 100). He also noted a range of military classifications of troop types: officer classes, cavalry, legionary and auxiliary soldiers, archers, slingers, *symmachiarii* (men

Fig 52. Decorative devices on semi-cylindrical 'legionary' shields from Trajan's Column (1). (Artwork by J. R. Travis, from Rossi, 1971; and James, 2004)

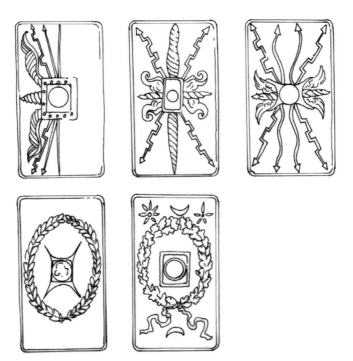

Fig 53. Decorative devices on semi-cylindrical 'legionary' shields from Trajan's Column (2). (Artwork by J. R. Travis, from Rossi, 1971; and James, 2004)

with barely any equipment or clothing), marines (*classiarii*) and other sailors/rowers (*gubernatores*; Rossi, 1971, 101–105). Within this range of troop types, he identified two basic shield shapes in use, rectangular (tile-shaped) and oval; the former he attributed generally to legionary/praetorian use, and the latter to auxiliary units (Rossi, 1971, 108). He then noted a range of decorative emblems and motifs which appear on these shields in different combinations, which he believed may identify the individual units, some features being predominantly legionary, others auxiliary (Figs 52–56; Rossi, 1971, 108).

THUNDERBOLTS

One of the most common symbols frequently seen on rectangular legionary/praetorian shields, and occasionally on oval auxiliary shields, is the thunderbolt and lightning, or *leitmotif*, formed by zigzagged flashes, radiating out from the *umbo*, terminating in arrowheads. On Trajan's Column, this symbol, signifying Jupiter's divine power (frequently used in other Roman contexts, including coinage, to represent Imperial supreme authority), is often associated with wings (representing winged victory), and may also be seen gripped in the claws of a legionary eagle (*aquila*). This use is not confined purely to Trajan's Column however, also appearing in other sculptural representations in legionary/praetorian contexts, such as on the Column of Marcus Aurelius and on the Adamklissi metopes, suggesting that these images at least were drawn from life and were not entirely symbolic or products of the sculptor's imagination.

It has been suggested that the thunderbolt-and-lightning motif signifies the status of Roman citizenship, a view with which Rossi (1971, 111) concurred where found in legionary/praetorian contexts. However, when found on oval shields of presumed auxiliary use, this interpretation fails. Rossi believed that here it may represent members of the Imperial bodyguard, or *singulares imperatoris* (as on auxiliary oval shields of the Imperial

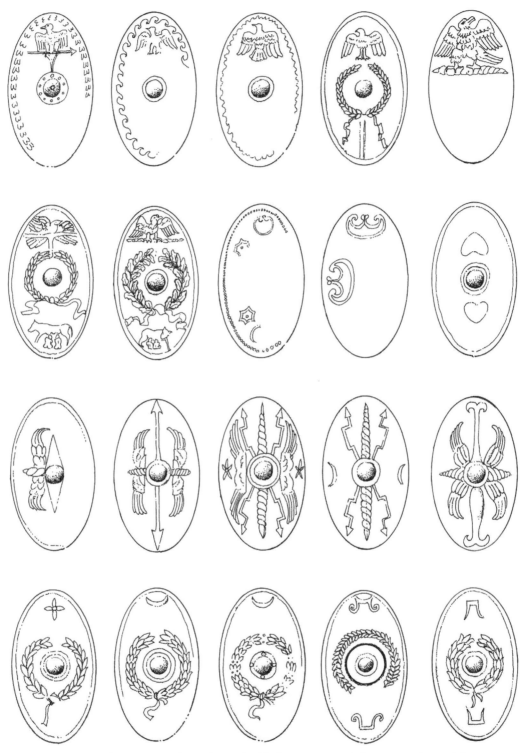

Fig 54. Decorative devices on oval 'auxiliary' shields from Trajan's Column (1). (Artwork by J. R. Travis, from Rossi, 1971; and James, 2004)

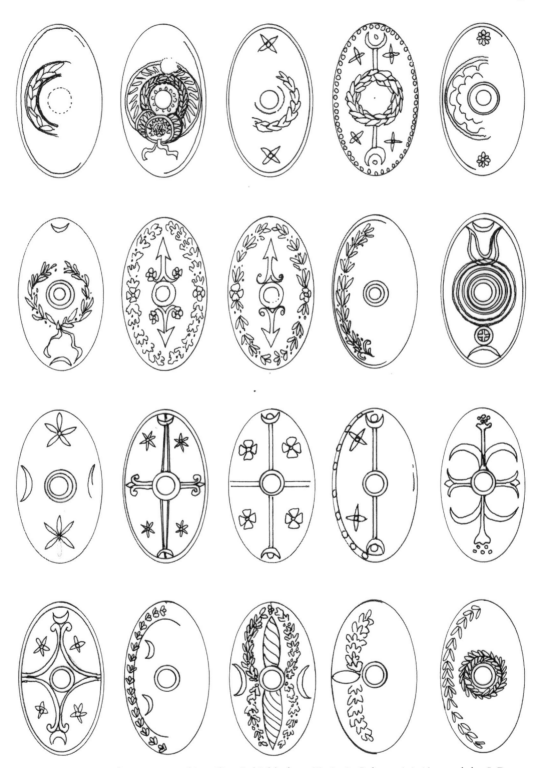

Fig 55. Decorative devices on oval 'auxiliary' shields from Trajan's Column (2). (Artwork by J. R. Travis, from Rossi, 1971; and James, 2004)

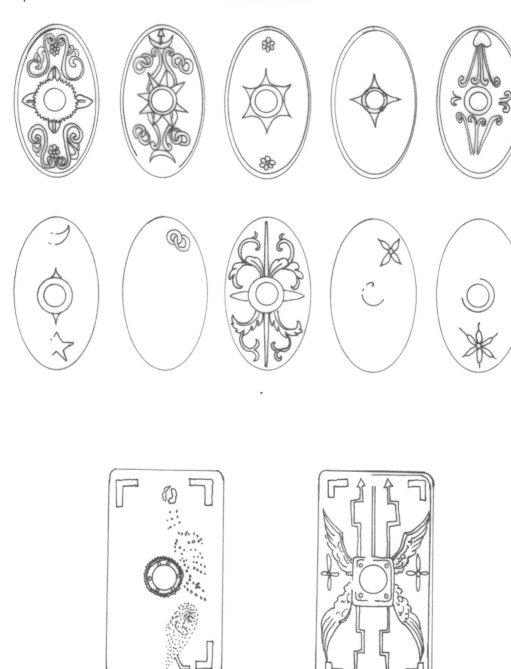

Fig 56. Decorative devices on oval 'auxiliary' shields from Trajan's Column (3). (Artwork by J. R. Travis, from Rossi, 1971; and James, 2004)

bodyguard depicted on the Domitianic relief, Rome), or of the *cohortes voluntariorum civium Romanorum* and *ingenuorum*. These were units attached to the emperor's person, and/or units drawn from men either originally possessing Roman citizenship or awarded citizenship for distinguished service (*ante emerita stipendia*; Rossi, 1971, 115).

Units which would fall into these categories include: *ala Imperatoris civium Romanorum*; *Imperatoris praetoria singularium*; *cohors Imperatoris campestris civium Romanorum*; *Imperatoris Cisipadensium*; *VIII voluntariorum civium Romanorum*; *Pedites singulares Britannici*.

Dando-Collins (2010, 77–81) also discussed the use of thunderbolts and how shield symbols changed over time. He echoed Rossi's view that the thunderbolt was representative of citizen troops (Rossi, 1971, 111), although he was more specific. In his opinion (Dando-Collins, 2010, 79), he considered that thunderbolts should be attributed to the Praetorian Guards, that they were the only ones who used them (contrary to the units listed above by Rossi). Dando-Collins' view was that the images on Trajan's Column are erroneous, as the artisans probably used Praetorian Guards as models and that, being city-based, they had probably never seen any military equipment other than Praetorian, although he did not appear to have considered

Fig 57. Caius of *legio II Adiutrix*. (Artwork by J. R. Travis, redrawn from Russell-Robinson, 1975, 167, plate 470)

the possibility of war artists accompanying the legions on campaign, sketching actual equipment. However, despite this assertion that the thunderbolt symbol was only ever used by Praetorians, later in his work, when listing the histories of each legion, he cited its use by several non-Praetorian legions (for example *II Traiana*; *XI Claudia*; *XII Fulminata*; *XIV Gemina Martia Victrix*; and *XXX Ulpia*; 2010, 116, 163, 165, 169, 190). It should also be noted that Caius, legionary of *legio II Adiutrix*, is depicted on his *stela*, resting his hand on an oval shield with a slightly dished shape, which is highly decorated with zigzagged lightning/thunderbolts (Russell-Robinson, 1975, 167, plate 470; Fig. 57).

Other possible indications of Roman citizenship (according to Rossi; 1971, 114), which are often seen, individually or together, are the eagle and the she-wolf symbols. Occasionally these also can appear on oval auxiliary shields, which Rossi (*ibid.*) considered was used to visually distinguish units who may have earned citizen status for valour, as with the *civium Romanorum*. The eagle would usually appear in the top half of the shield, perched on a thunderbolt or arrow, and the she-wolf, protecting or suckling her twin human babies, would appear in the lower position.

Another design, appearing on both rectangular legionary/praetorian and oval auxiliary shields, is a circular crown or wreath of leaves, centred on the *umbo*, which Rossi interpreted under two different categories, *coronae* and *torques* (Rossi, 1971, 113). The *coronae*, more commonly seen on legionary/praetorian shields, Rossi considered comparable to civic, triumphal crowns handed to citizens in triumphal parades – and these appear on the shields of units who have distinguished themselves in action, for example those of *legio XXX Ulpia*, the first soldiers to cross the Danube (*ibid.*) – and also suggested that similar *coronae* may appear on shields belonging to units of auxiliaries associated with these legions (Rossi, 1971, 114). Rossi felt that the *torques*, or wreathes of leaves, more frequently seen on auxiliary shields should be interpreted differently to the legionary *coronae*, comparing them to festoons seen on altars, described by Virgil as *nexis ornatae torquibus arae* (*Georg.* 4.276), and to the circular wreaths used as a cohort emblem at the apex of its *signa*, or standard. Rossi compared this also to the ornamental twisted collars worn by Celtic/barbarian people as symbols of honour, translated visually onto the shields of auxiliary troops rather than being physically worn in barbarian fashion (Rossi, 1971, 116). These units, in addition to the honour of visual *torques* appearing on their shields, would also be bestowed with corresponding titles of *torquata*, *bistorquata* (double wreathed), or *torquata armillata* (representing bracelet/arm torques).

These *coronae* or *torquata* are also occasionally seen with a ribbon underneath, being seen as representative of cohorts on their *signa*. They are also sometimes associated with other star-shaped or crescent-shaped symbols, or animal motifs, which may be representative of a specific legion, such as the lion, seen on the Arch of Galerius (Russell-Robinson, 1975, 93, Fig. 123) and again on the Dura Europos shield (James, 2004, Plate 10), and the bull, as on the South Shields shield boss (Bidwell, 2001, 10, Fig. 5A).

Some of the oval auxiliary shields on Trajan's Column were notable for their lack of any Roman-style decoration, for example those belonging to the lightly armed *symmachiarii* (who lacked any possibility of gaining future citizenship), using very simple motifs of stars, interlaced rings or ornamental volutes. Rossi further identified two rectangular legionary-style shields with no Roman emblems. These, he suggested, may belong to disgraced units which had been deprived of honours for cowardice or disloyalty, which had since been disbanded and had suffered *damnatio memoriae*, their names and symbols erased, replaced with swirling, volute patterns, as for example with the *legio XXI Rapax* (Rossi, 1971, 112), or as seen on the sculptural relief of the god Mars from the votive pillar at Mavilly, Cote d'Or (Fig. 58).

SHIELD EMBLEMS USED AS LEGION-SPECIFIC IDENTIFICATION

Dando-Collins (2010, 61–64), in his work on the 'Legions of Rome', discussed in greater depth the use of legion-specific emblems found on shields (along with other identification markers, such as tile stamps). He noted (2010, 77) that the most frequent emblems seen were animals or birds (especially those with religious significance, such as the eagle, bull, stork or lion), with some representations from Greco-Roman mythology (such as Pegasus, the Centaur, Mar's thunderbolt and Neptune's trident). As a means to explain some of the reasoning behind the symbolism, he used Lawrence Keppie's legion number formula to discuss the origins of the legions, in particular of the fifth to tenth legions.

In his discussion (2010, 61–64, 77–81), he further suggested that some of the emblems should be seen as zodiacal birth signs representing the original foundation of the legion,

Fig 58. Relief of god Mars from votive pillar at Mavilly, Cote d'Or. (Artwork by J. R. Travis, from Russell-Robinson, 1975, 169, fig. 176)

with other signs denoting its recruitment area. For example, he discounted the previous suggestion that all legions using the bull emblem had been raised by Caesar and those using Capricorn (the sea goat) by Octavian. He asserted, however (*ibid.*), that the majority of Caesar's legions did not use this symbol, (Caesar's personal symbol being the elephant). He further noted (*ibid.*) that of the four legions raised by Caesar in Italy in 58–56 BC (the 11th to 14th), none used the bull, but three of Octavian's legions not connected to Caesar did use the bull (see Tables 3, 4 and 5).

The Bull
In his argument, Dando-Collins (2010, 61) cited Livy (39.30.12) that the fifth, seventh, eighth legions were stationed in Spain in the second century BC (in 185 BC the fifth and eighth legions; in 181 BC the fifth and seventh legions). Previously, the eleventh, twelfth and thirteenth legions had also been campaigning in Cisalpine Gaul. According to Keppie (1984, 2), legion numbers 1 to 4 were reserved for the Consuls, based in Italy; legions 5 to 10 were in Spain; 11 to 13 in Cisalpine Gaul; and higher numbers towards the east. In 61 BC, when Caesar was governor of Baetica (Further Spain), he reported that legions 5, 6, 7, 8 and 9 were based in Spain, Caesar then raising a sixth, the tenth legion. Caesar then wrote (*Bell. Gall.*, 1.24) that in 58 BC he was served in Gaul by four of these 'veteran legions'. Keppie suggested that this was probably the seventh, eighth, ninth and tenth legions, with the fifth and sixth legions remaining in Spain. The eleventh and twelfth legions were also raised around this time in Cisalpine Gaul.

Three of the consular legions (second, third and fourth), plus another unnamed (possibly the *Martia*), were then sent to Spain (to replace those transferred to Gaul) bringing the total back up to six legions. These were under the control of Pompey the Great, but on the surrender of his forces in Nearer Spain in 49 BC they were discharged and sent home. Caesar (*Bell. Civ.*, 1.86) wrote that a third of these returned home to Spain, and Dando-Collins (2010, 63–64) proposed that they were men of the fifth and sixth legions, originally recruited in Spain. This, he then suggested, supports Keppie's view of legions 5 to 10 being founded and recruited in Spain, no legion numbered over 10 ever being stationed there. He further noted that all of these traditionally used the bull emblem, concluding that the bull should therefore be interpreted instead as denoting legions recruited in Spain, possibly linked to the ancient tradition in Spain for bullfighting. He also acknowledges some exceptions, as when other legions occasionally adopted different symbols (as was also sometimes reflected in their names), as representative of notable victories and achievements. For example the *legio V Alaudae* ('The Larks') took Caesar's elephant symbol after their success in defeating King Juba's elephants in the 46 BC Battle of Thapsus, although prior to this they had probably also used the bull symbol. Similarly, the *legio III Gallica* used the bull, but served in Spain under Pompey from 59–49 BC, and *IV Flavia*, who replaced *IV Macedonica*, took the Flavian lion emblem (this being the emperor's family symbol).

Fig 59. Shield from Arch of Orange attributed to the *legio II Augusta*. (Artwork by J. R. Travis, from Russell-Robinson, 1975, 169)

The Boar
Dando-Collins (2010, 78) reported that the Celts traditionally used the boar as a symbol to ward off evil. He considered that Celtic

tribes living in Cisalpine Gaul, who had been incorporated into Italy in 42 BC, would have retained some of their traditional customs. He therefore thought it reasonable that several legions raised in Italy would have used the boar (for example *I Italica* and *XX Valeria Victrix*). There is, however, also evidence from Etruscan wall paintings at Tarquinii that the image of a boar was already a well-established symbol found on shields dating from the fourth century BC.

The Centaur
This symbol is usually associated with Thessaly in Greece, so therefore Dando-Collins (2010, 78) again considered it a reasonable choice to use for the three legions raised in Macedonia and Thrace at the end of the second century AD (e.g. first, second and third *Parthica* legions).

Capricorn (the Sea Goat)
The suggestion had been made that all legions using Capricorn had been raised by Octavian. Dando-Collins (2010, 78), however, disagreed. He cited examples of many legions raised after the time of Octavian/Augustus, who had still been using the Capricorn symbol (such as *I Italica, I Adiutrix, II Italica, II Adiutrix, XXII Primogeneia* and *XXX Ulpia*). He considered that it was instead a zodiacal birth sign, denoting the time when the legion was first raised. Capricorn was the sign for midwinter, which was the time when most legions were traditionally raised for the next spring, so as a consequence appears to be the sign most frequently used (for example as seen on a shield from the Arch of Orange, attributed to a Centurion of *legio II Augusta*; Fig. 59; Russell-Robinson, 1975, 26, Fig. 31).

DECORATION ON THE DONCASTER SHIELD

Buckland (1978, 251) identified fragments on the boss, handgrip and areas of charred wood as being remains of a leather covering on the Doncaster shield. Numerous small iron studs were found (solid domed heads 10 mm in diameter by 6 mm high, with square tapering shafts 3 mm wide by 113 mm long), hammered in from the front, some close enough together to touch heads, and clenched on the back, which Buckland (1978, 253) interpreted as securing the leather cover, forming four diagonal lines radiating from the boss. One row of nine undisturbed studs was found, others were found scattered, and one small group he interpreted as forming a 'lunate' shape.

At the northern end of the remains, the upper end of the *spina*, the thin wooden midrib (33 mm wide by 5 mm high) of curved cross-section, on the front of the shield, still preserved part of its sheet bronze covering, secured by small, short bronze studs, some with flat and some with domed heads (Buckland, 1978, 253). Buckland also reports fragments of sheet bronze decoration on the rest

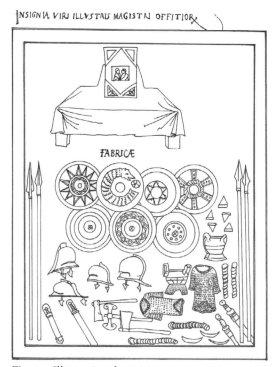

Fig 60. Illustration from *Notitia Dignitatum: insignia viri illustris magistri officiorum*, from sixteenth-century copy of fifth-century manuscript. (Artwork by J. R. Travis)

of the shield, in parts fixed down by the *spina* and its studs. Some small fragments of sheet bronze decoration with curved fractures were interpreted by Buckland as being evidence of chased curvilinear patterning on the sheet, and the fact that little of this decoration remains is interpreted by Buckland as being due to its being stripped off before burning (Buckland, 1978, 256).

DECORATION ON THE DURA EUROPOS SHIELDS

Although similar in shape to the typical legionary semi-cylindrical shields of the first and second centuries AD, the decoration used did not feature the traditional 'winged thunderbolt' seen on the majority of shields depicted on Trajan's Column, as seen on the legionary tombstone of Gnaeus Musius at Mainz. However, another decorative feature commonly occurring on Trajan's Column was the wreath, or *corona*, and although seen as circular in form when around circular bosses, here similar concentric borders of wreathlike leaf patterns can be seen around the apparent rectangular boss placement area (Figs 52–53).

James (2004, 164) also notes the absence of the 'winged thunderbolt', suggesting that it may have disappeared from fashion as a motif from the late second century AD, and was in any event completely missing from the late fourth century AD designs as represented on the *Notitia Dignitatum* (Fig. 60). Other comparatives to Trajan's Column can be drawn, however, in the eagle figure (*aquila*) located above the wreath. This is seen on Trajan's Column, also associated with a she-wolf figure below the wreath, which is similar in format to the Dura example, which substitutes instead a lion on the lower position. Again, this echoes in format a similar shield on the Arch of Galerius, also with a lion in this position, and another on the Piazza Armerina mosaic with a boar. In his study of the shield motifs of Trajan's Column, Rossi saw this *corona* design, with the legionary *aquila*, as a better emblem for Roman citizenship than the 'winged thunderbolt' motif (Rossi, 1971, 112), with the increasing incidence of acquired citizenship awarded for valour to units of *voluntariorum civium Romanorum* and *ingenuorum*. Therefore, it is perhaps not surprising that this symbolism supersedes in later periods, the animal motif in the lower position perhaps being legion specific.

This format is again similar to that seen on the decoration of the rectangular shield boss from South Shields, where the *aquila* takes central position within a wreathed *corona* on the *umbo* itself, and a bull, symbol of the *legio VIII Augusta*, takes the lower position (Fig. 44; Bidwell, 2001, 10, Fig. 5A). In addition, all of these eagle/wreath/animal designs, the Dura example being no exception, are also often associated with lesser motifs of stars and crescents (James, 2004, 164; Rossi, 1971, 109).

CHANGES TO SHIELD SYMBOLS IN THE LATER EMPIRE

As previously stated, James (2004, 164) had commented on the absence of the 'winged thunderbolt', beginning possibly in the late second century AD but in any event established by the late fourth century AD, as represented on the *Notitia Dignitatum* (a view repeated by Dando-Collins; 2010, 80). The latter remarked that he would have expected to see some Christian symbols, but found very few crosses and no designs using the 'XP' (chi-rho) symbol, although these had been ordered to be painted on shields by the emperor, Constantine (Dando-Collins, 2010, 80). He did however, note the presence of the 'wheel' symbol of the pagan goddess, 'Fortune', citing references in Ammianus Marcellinus (31.1.1), to 'Fortune's rapid wheel', also associated with the war goddess, 'Bellona'. The earliest evidence for this wheel symbol on shields dates back to the first quarter of the first century AD: for example, a terracotta sarcophagus from Vacano shows two shields decorated with geometric devices forming a wheel design.

By the fifth century AD, he noted that the 'bull' appears to have been replaced by a 'rosette' (*Notitia Dignitatum*), which was also a war symbol associated with *Bellona*, being seen on shields and grave *stelae* from the early Imperial period (2010, 80). Also by the fifth century AD, other legions had similarly been seen to have replaced their emblems: *III Augusta* (to a plain circular design); two Imperial seventh legions (one with a ten-pointed star; one with a nine-spoked wheel of fortune); *I Italica* (who replaced the boar with a circular motif); *II Italica* (with four-spoked wheel). Some continuity was also noted however, for example with the *XIII Gemina*, who were still found to be using the lion emblem (*ibid.*).

REMOVABLE SHIELD COVERS

As discussed previously, shields would have been covered by layers of woven cloth or leather, and their surfaces sealed by the application of a protective coating of gesso and paint. The painted surface would have not only waterproofed the structure, lengthening its period of usefulness, but would also have served as an identifying feature, to enable fighting units to easily differentiate allied from enemy combatants (as described in contemporary accounts from literary sources that the shield board itself would also have been decorated as a means to identify troop units; Ammianus, 16.12.6; Claudian, *Bel. Gild.*, 423; Vegetius, 2.18). Furthermore, the evidence from remains of shields from Dura Europos, along with artistic representations on funerary *stelae* and monumental sculptures (such as the Arch of Orange and Trajan's Column), suggests that this decoration may have exceeded just the basic single-colour application, providing symbols denoting not only specific legion (perhaps symbols of both legion identity and 'birth' date), but also perhaps of subdivision within that legion by slight variances of design layout and shield shape.

However, in addition to this painted protection, it is apparent from both literary and artefactual evidence that a further layer of protection was afforded by way of removable waterproof covers. Both Caesar (in his *Gallic War*) and Cassius Dio (*Roman History*) described the use of shield covers. Caesar (*Bell. Gall.*, 2.21.5) suggested that covers were made in such a way as to be quickly removed if needed, but also that they would not greatly impede their use if covers remained in place (noting one occasion of Gauls attacking so unexpectedly that they were not able to remove the covers but were still able to use their shields).

However, this was apparently only the case under favourable weather conditions, with leather covers only able to withstand light rain, becoming heavy and unwieldy under severe downpours. In contrast to Caesar's description, Cassius Dio (56.3) described how the legions of Varus were attacked suddenly under heavy rain, weighing down the soldiers with their

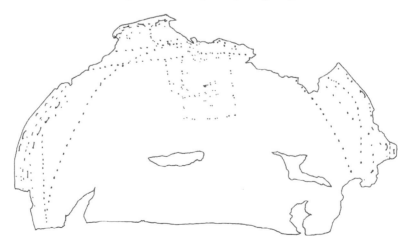

Fig 61. Removable covers for shields (as found at Caerleon). (Artwork by J. R. Travis, from Van Driel-Murray, 1988)

wet clothing and equipment, and how the shield covers and shield surfaces became swollen with water, to such an extent that they were unable to remove the covers (Van Driel-Murray, 1988, 53, also presumed that this may have also warped the ply structure of the boards).

Remains of a number of these removable leather covers have been found at several European sites, where waterlogged deposits have allowed for a greater level of preservation of perishable leather remains. Van Driel-Murray (1988, 51–65; 1989, 18–19; 1999, 45–54), has made detailed study of several goatskin shield covers, including those from Caerleon (Fig. 61) and Castleford (both in Britain), and Roomburgh (near Leiden, Netherlands; Fig. 62), described in greater detail below. She reported similar covers also being found from Valkenburg (Fig. 63; Groenman-van Waateringe, 1967, Fig. 17), and Vindonissa (Fig. 63; van Driel-Murray, 1999, 45–54), while Feugère (2002, 93) and James (2004, 169) report further fragments from Birdoswald, Hardknott, Vechten and Velsen. James (2004, 169) considered that despite no such covers being found at Dura Europos, they may have been used there, although the evidence had not survived, with the possibility of the use of textiles as an alternative to leather, but with a lower level of preservation.

REMAINS OF COVERS

In her studies of the Caerleon and Roomburgh shield covers (1988, 51–65; and 1999, 45–54), Van Driel-Murray compared the remains of several rectangular and oval-shaped

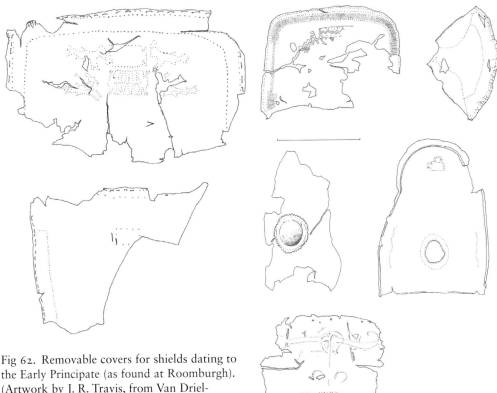

Fig 62. Removable covers for shields dating to the Early Principate (as found at Roomburgh). (Artwork by J. R. Travis, from Van Driel-Murray, 1999)

Fig 63. Removable covers for shields dating to the Early Principate, as found at Vindonissa (rectangular and central shield boss area); Castleford (circular); Valkenburg (oval). (Artwork by J. R. Travis, from Van Driel-Murray, 1999; & 1989)

legionary and auxiliary shields, their sizes, the materials used in their construction, and the method of their attachment. She also drew analogies to the smaller, circular shield cover from Castleford (1989, 18–19), which exhibits many similarities in construction, although it appears to belong to a more specialist shield type (e.g. for *signifier* or cavalry, sometimes referred to as a *clipeus*; Fig. 6), as seen on Trajan's Column (the shield size of 48–50 cm proposed by Van Driel-Murray as 'fitting nicely between armpit and wrist').

During excavation in 1996 of the Roman fort at Roomburgh (near Leiden, Netherlands), two large fragments of shield covers, along with some fragmentary pieces of appliqué decorations, were found in a bundle of other scrap leather remains (ten tent panel pieces), which had been discarded into a ditch associated with the fort's *vicus* (Fig. 62; Van Driel-Murray, 1999, 45). The pieces of leather in the bundle had been presumably cut from worn or broken equipment and retained for later reuse. The level of usage wear and the style of seams used suggested the estimated age of manufacture as the end of the first to the start of the second century AD, with deposition probably in first two decades of the second century AD (*ibid.*).

Both of the Roomburgh covers appear to have been intended for large, rectangular shields, although, being incomplete, their full height is not known. Cover 1 appears to be around 82 cm wide, and although Cover 2 is less complete, based on their similarities it was presumed to be of similar width (Van Driel-Murray, 1999, 47). Both covers were edged with a folded drawstring hem, with an additional 4 cm-wide strip sewn parallel to the edge. This was interpreted as serving to reinforce the edge when pulled around the shield (*ibid.*). Van Driel-Murray also noted that an impression of the width of the board edge was visible on the inside (flesh side), suggesting a board thickness around 10 mm, further estimating that the intended board would have been around 75 cm wide, with 4–5 cm of the cover turned over at the back when drawn up (*ibid.*).

In comparison, the fragmentary shield cover from Caerleon, found in 1985 in test pits, from similarly waterlogged deposits, appears to be contemporary with the Roomburgh examples, being dated by pottery to the first and early second centuries AD (Fig. 61; Van Driel-Murray, 1988, 51). The fragment was found to be a semi-circular piece of goatskin, 76 cm by 40 cm, which appeared to have been from an end part of a broad oval shield cover. The piece showed no signs of rivet holes for the attachment of any metal edging strips, which suggested that it had formed part of a removable cover, rather than having been glued permanently to the surface of the wooden ply base (Van Driel-Murray, 1988, 57). At the apex, the cover bore evidence of holes for decorative stitching and attachment of unit insignia, suggesting that this had come from the top section of the cover (*ibid.*). Stitching was evident only from an edging hem, not from any joining seam, with no other piece attached to it (Van Driel-Murray, 1988, 52). It appeared to have been roughly cut away from the rest of the cover, folded into three to set aside for possible reuse (similarly retained large pieces of previously used leather are frequent finds from Roman military contexts, where it must have been common practice to squeeze the maximum use out of any resources).

Puckering of the leather in the turned hem, forming a 'channel' shape, has been proposed to have resulted from the use of a drawstring cord used to pull the cover tight around the shield, which would have emerged at four small, reinforced holes in the centre hem at each side (as is seen in the Valkenburg cover). Tearing at the lower left and right corners of the Caerleon fragment has been interpreted by Van Driel-Murray (1988, 52) as possibly being caused by the pulling out of these drawstring ends.

A similar method of drawstring attachment appears to have been used on the Castleford cover. The fragment appears to form a quarter of a circular leather cover of diameter around 60 cm, reduced to around 56 cm by its edging hem. The outer edge has again been turned over, folded and tacked, probably for use with a drawstring. Allowing then for a turnover fold of 3 cm to draw around the circular shield, this would suggest a board size of 48–50 cm diameter (Van Driel-Murray, 1989, 18–19).

As the Caerleon fragment had been cut away on its lower edge, not reaching its original full extent, it was not possible to define the position of the *umbo*, and so estimate the full length of the cover (Van Driel-Murray, 1988, 57). No *umbo* cover remains but Van Driel-Murray estimated its size as around 15 cm diameter, possibly round in shape, a cross between a 'cone' and a 'hemisphere', stitched along the side and moulded into shape, this then being stitched in place on the cover with an additional strengthening ring of leather placed around it (Van Driel-Murray, 1988, 63).

The direction of the grain suggested that a full length of goatskin had been used, laid horizontally across the width of the shield, with the maximum width of a goatskin taken as around 50–55 cm (evidence of pre-tanning processes, 'fleshing' and 'scudding' being noted on the 'flesh' side; Van Driel-Murray, 1988, 52). Van Driel-Murray (1988, 57) therefore proposed that the full cover would require two skins, seamed across the middle at the *umbo*, as on the large oval cover from Valkenburg (Groenman-van Waateringe, 1967, Fig. 17). Similarly, the goatskins used in the Roomburgh covers appeared to have also been arranged horizontally. In these cases, however, as there were no signs of *umbo* position on the panels, this was assumed to have been on the middle panels, with the probability of three skins being required, producing shields of greater length (Van Driel-Murray, 1999, 47).

DECORATION OF COVERS

As reported in his *Roman History*, Cassius Dio (56.3) described how shields could become unusable if weighted down by their covers when soaked in heavy downpours. Nevertheless, the presence of these leather covers suggests that they must have offered some level of weather protection. This further suggests some form of waterproofing must have been applied to the leather. As with the shield board, this could have been painted surface decoration, but this would be most unlikely. Painted coating of the shield cover with yet another easily damaged layer of decoration would defeat the whole purpose of covering the original painted board surface. Furthermore, the remaining cover fragments found to date do not exhibit any retained traces of painted surface decoration. It is possible that dyes and stains may have been used; again no traces have been found, although these may have been leached out during their time submerged in the waterlogged deposits.

Van Driel-Murray (1988, 57) assumed that vegetable-tanned leather had probably been used because of the level of preservation, although she felt unable to say what tannin agents may have been used, as tannin would have been naturally present from rotting organics within the waterlogged deposits. In her description of the reconstruction made of the cover, Van Driel-Murray (1988, 57) reported that rough, undyed, vegetable-tanned goatskin had therefore been used, heavily oiled with cod oil as a waterproofing agent (although there is no solid evidence for this being used), with decoration being defined by use of differently shaded natural leathers and contrasting coloured stitching. Similarly, stitching holes for appliqué design were found on the circular shield cover from Castleford, with a symmetrical arrangement of 'petals' placed over a semi-circular panel, the decoration possibly being enhanced by the use of different-coloured leathers, which, although not serving any obvious purpose of identification, would have provided extra protection of the shield edge (Van Driel-Murray, 1989, 18–19).

In the case of the Caerleon cover, lines of awl holes can be seen, where a continuous running stitch had been used (using two threads passing simultaneously from each side), which Van Driel-Murray (1988, 53) interprets as showing where applied panels were sewn to the front. She further interprets curved stitching on the body of the cover as being for an assumed attachment of a larger panel, oval in shape. She also speculates that this panel may have been of cloth, perhaps painted like the Dura Europos shield boards, used as backing support for shield designs (*ibid.*).

The military unit's identification symbol would probably have been attached at the top, although no thread impressions could be identified on the flesh or grain side. This Van Driel-Murray (*ibid.*) interprets as where an appliqué design would have been sewn onto backing material and whip stitched to the panel, as on covers from Vindonissa, with the suggestion that a second separately applied piece may have also been sewn to the centre of the panel. Stitching to pick out designs on an applied panel is also seen on covers from Valkenburg and Bonner Berg, but in the case of the Caerleon cover it was not sufficiently detailed to be able to identify the unit (*ibid.*). In view of the crudity of the method of attaching the insignia, she therefore proposed that this could be interpreted as indicative of professional, centralised manufacture of 'blank' covers.

While it was not possible to identify the unit owning the Caerleon shield cover, other than by location and estimated date of use, in contrast, both of the rectangular shield covers from Roomburgh bore traces of appliqué panels (toffee wrapper-shaped *ansata*) to show unit designation (that for Cover 1 being clearly marked with the owner's unit identity, although in the case of Cover 2 this unit was unknown) at presumed top-centre position (van Driel-Murray, 1999, 47, figs 2 & 3).

The appliqué of Cover 1 has similar decoration to Cover 2 but is more elaborate, identifying the owner as being a member of the '*Cohors. XV Voluntariorum Civium Romanorum*' (Van Driel-Murray, 1999, 47). Located just below the edge reinforcement, at top-centre position, stitching outlines can be seen for three appliqués: one for *ansata*, flanked by two others, shaped like Capricorns. These, however, are simple outline shapes, with no internal decoration, which must have been stitched, carved or painted on the appliqués directly (Van Driel-Murray, 1999, 47).

These three appliqués appear to have been later replaced by more elaborate versions, for example showing more detail on the Capricorns, in addition to a possible star or comet motif above them, and featuring the identifying lettering of 'COHR/XVVO'. This was identified by Van Driel-Murray (*ibid.*) as referring to '*Cohors XV voluntariorum civium Romanorum pia fidelis*', who were known to have been in the area around AD 200. She further proposed the possible reason for the user having changed the original appliqué as being to remove the lettering for the honorary award of 'P. F.' (granted AD 89) due to Domitian's later '*damnation memoria*' (Van Driel-Murray, 1999, 47).

Also found with the covers were a number of fragments of goatskin appliqués (ivy leaves, hearts, pieces of scalloped frame), which would have been stitched to a backing piece, so that a larger appliqué panel could be more loosely tacked into position (Van Driel-Murray, 1999, 49). Unable to identify if the design would have been a decorative pattern or further lettering, Van Driel-Murray (*ibid.*) drew analogies to similar decorative panels (for example as at Qasr Ibrim, Egypt) and on other covers (such as from Valkenburg, Vindonissa, and at Caerleon as described above). She proposed the use of *ansata* panels, with the inscriptions cut out or stitched in, the appliqués then stitched down, using visibly coloured twine, or contrasting coloured leather ('untanned', rather than 'tanned', as used for covers) as this better accepts coloured 'stains' (but noting that these do not survive so well in waterlogged conditions as providing a reason for lack of remains).

SHIELD SHAPES, AS SUGGESTED BY REMAINS OF REMOVABLE COVERS

In her discussion of the size and shape of the shield that the Caerleon cover was intended for, Van Driel-Murray (1988, 57) assumed its proportions to have been probably the same as for the broad, oval cover from Valkenburg, but at 149 cm by 80 cm being considerably larger, which she considered to be 'impossibly large'. She proposed that the shield should have been larger than that from Valkenburg, but not of the same proportions (being shorter in relation to its width), so perhaps with a length of 120–125 cm (that being in the same length range as those from Doncaster and Dura Europos. However, in her discussion of the Roomburgh covers, she further suggested that variation in shield length may simply relate to the average height of the men in a unit (providing for a range of 102–128 cm; Van Driel-Murray, 1999, 51/52).

Although it was not clear how much of the Caerleon cover should have been drawn over the shield edge, in her view it may have been oval in shape, straight sided with a curved top; in the reconstruction by J. Pickett-Baker being interpreted with straight-line stitching on the sides and across the top, with breaks at corners (Van Driel-Murray, 1988, 59–60, figs 4 & 5). However, the stitching holes shown in the illustrated remains in the report (Van Driel-Murray, 1988, 54, fig. 1), do seem rather more indicative of a shape more like that of the Doncaster shield: straight sided, with curved corners and only a gently curved top and bottom (almost straight). It is also possible that the cover could even be for a shield with both a straight top and sides, with corners of the cover not overlapping each other, and so reducing any bunching up of excessive gathered material at the corners.

In her report on the rectangular shield cover of the *Coh. XV Voluntariorum C.R.*, Van Driel-Murray (1999, 49) discussed the shape and size of a number of covers, particularly whether the shields would have been oval or rectangular in shape, noting that the majority of covers found have been for oval shields. She then offered a table of cover types/board types (1999, 51/52); with small oval types and pointed oval types as from Valkenburg seen as auxiliary shields, but excluding the particularly wide Dura Europos shield (now included in the revised Table 6 below), its presented width to an adversary being, however, reduced by its greater curvature around the user. Within the remaining shields she proposed three sizes of cover for larger shields (excluding the specialist, circular *signifier*'s shield from Castleford, described above, and listed along with the Dura shield at the end of the table for comparative purposes), these being: i) broad oval; ii) rectangular (both being for boards around 64 cm wide); and iii) for larger shields around 75 cm wide, these including the two Roomburgh covers. It is also particularly interesting to note here that she then proposes these larger shields then to be similar in shape/size to the Doncaster shield, which she describes as 'lightly curved' (Van Driel-Murray, 1999, 51/52), as has also been proposed here in previous chapters.

In her discussion of the shape of the Roomburgh covers, Van Driel-Murray (1999, 52) proposed that the broad, oval shield shape may not have been 'an early second-century phenomenon'. She noted (*ibid.*) that rectangular shields are seen on Trajan's Column, but that there is an absence of rectangular covers from the Hadrianic period (based on Hadrianic-period *fabricae* waste from Bonn), concluding that a widespread use of rectangular shields may have been confined to a relatively short period, from mid-first century to first decade of second century AD (the dating range of the Roomburgh covers).

Van Driel-Murray (1999, 53), argues that in the naming of the *Cohors XV voluntariorum civium Romanorum*, the term '*voluntariorum*' indicates that its members were citizen volunteers (including some ex-legionaries), and consequentially that they would have been kitted out as legionary troops (associating the rectangular shields with legionary use). She proposed that the unit may have been stationed in the area under the rule of Domitian (as discussed, probably at the time when the unit lost its honorary award of '*pia fidelis*'), and/or Trajan, and that the dumping of the covers may indicate a change in equipment in the early Hadrianic period when, in her view, the rectangular shields may have fallen from use.

UNIT	EMBLEM	BIRTHSIGN	FOUNDATION	RECRUITMENT
1st ADIUTRIX	Pegasus	Capricorn	AD 68	Gallia, Italy, Narbonensis
1st GERMANICA	Possibly Pompey's Lion with sword in paw	Capricorn	Republican	Orig. Italy, later Spain
1st ITALICA	Boar	Capricorn	AD 66	Italy
1st MINERVIA	Poss. Gorgon's Head, symbol assoc. with Minerva & used by Domitian	Aries (ram)	AD 82	Provinces
1st PARTHICA	Centaur	Capricorn	AD 197	Macedonia & Thrace
2nd ADIUTRIX PIA FIDELIS	Pegasus	Capricorn	AD 69	Gallia, Italy, Narbonensis
2nd AUGUSTA	Pegasus	Capricorn	Republican	Northern Italy
2nd ITALICA	She-wolf & twins	Capricorn	AD 165	Originally Italy
2nd PARTHICA	Centaur	Capricorn	AD 197	Macedonia & Thrace
2nd TRAIANA	Hercule's Hammer & Lightening bolts	Aries (ram)	c. AD 105	German provinces
3rd AUGUSTA	Lion (from 1st C. coins), possibly also Pegasus	Capricorn	Pre-30 BC Octavian	Orig. Cisalpine Gaul, later North Africa
3rd CYRENAICA	Probably Ammon/ Jupiter (horned)	Capricorn	c. 36 BC	Province of Cyrenaica
3rd GALLICA	3 Bulls	Capricorn	49 BC	Orig. Gaul, later Syria
3rd ITALICA	Stork	Capricorn	AD 165	Italy
3rd PARTHICA	Centaur	Capricorn	c. AD 197	Thrace/ Macedonia

Table 3. Military units – shield emblems and birth signs (1) (from Dando-Collins, 2010).

Plate 1. Early Roman (Etruscan influence) 'hoplite' soldier of early fourth century BC (of Class I in the Servian system; Goldsworthy, 2003; see Table 9), wearing bronze Chalcidian-type helmet (which shows Greek influence); bronze cuirass tied at the shoulders; leg protectors (greaves), carrying small, round *hoplon* shield. (Artwork by J. R. Travis)

Plate 2. Samnite warrior of early fourth century BC, wearing triple-isk cuirass, carrying long, rectangular shield (*scutum*) and projectile weapons (javelin and prototype *pila*). (Artwork by J. R. Travis)

Plate 3. Celtic re-enactor and author
Will Llawerch portraying Iron Age/
Celtic warrior (wearing Celtic-style
mail with 'cape' shoulder doubling, as
seen on early first-century BC statue
of Mars from Mavilly, Côte d'Or)
with reconstruction of Celtic shield.
(Photograph by H. Travis)

Plate 4. Re-enactor portraying
legionary of first century AD.
Reproduction illustrates level of body
cover from curved, rectangular shield.
Decoration is a loose interpretation
of design from Arch of Orange,
attributed to *legio II Augusta*.
(Photograph by H. Travis)

Plate 5. Republican legionary (*Triarius*), from the start of the Second Punic Wars (218-201 BC), wearing bronze Montefortino helmet; long, sleeveless mail cuirass with shoulder doubling; leather arm guard (*manica*) and bronze leg protectors (greaves); carrying an early form of Roman short sword (*gladius*); and long, curved, oval Gallic-style shield. (Artwork by J. R. Travis)

late 6. Early reconstruction of legionary, wearing Newstead-style *lorica segmentata*, carrying legionary
hield, as seen on Trajan's Column. (Reconstruction made by H. Russell-Robinson, as displayed in Grosvenor
luseum, Chester in 1968)

Plate 7. Re-enactor portraying an auxiliary soldier of the Imperial period, first to second century AD (a member of Second Cohort of Tungrians), wearing mail cuirass (*lorica hamata*) with shoulder doubling closed by serpentine-style closure hooks, and helmet of Imperial Italic type D (as dated by H. Russell-Robinson to *c*. AD 83), carrying an 'auxiliary'-style oval shield. (Photograph by J. R. Travis)

ate 8. First-century standard bearer (*vexillarius*), carrying small, round parade shield. (Artwork by
R. Travis)

Plate 9. Legionary from marine unit from first century AD, wearing blue tunic and padded linen armour, carrying transitional semi-rectangular scutum with curved sides (painted blue and decorated with dolphins and Neptune's trident to denote marine unit). (Artwork by J. R. Travis)

Plate 10. Auxiliary cavalryman of Imperial period, wearing mail cuirass and 'Sports' masked face helmet, carrying oval cavalry shield. (Artwork by J. R. Travis)

Plate 11. Late Empire legionary of third century AD, wearing loose-fitting, long-sleeved scale cuirass, scale coif and copper alloy leg greaves (as depicted in the 'Battle of Ebenezer' fresco from the Synagogue, Dura Europos, and on the Arch of Diocletian, late third to early fourth century AD); carrying wide, oval convex shield, with round bronze boss (shield shows signs of combat damage and riveted bronze repair patch). (Artwork by J. R. Travis)

Plate 12. Late Empire centurion of fourth century AD, wearing thick woollen over-tunic over mail shirt. Geometric symbol on tunic hem is a symbol of centurion rank. His iron helmet is of Spangenhelm type (similar to that found from Der-el-Medineh, Egypt). He carries a wide oval shield, of gently convex shape, its decoration based on an image from the *Notitia Dignitatum*. (Artwork by J. R. Travis)

Plate 13. Re-enactor (J. R. Travis) wearing *lorica segmentata* (Kalkriese variant) of early first-century AD Roman legionary, carrying curved, rectangular *scutum*. (Photograph by H. Travis)

Plate 14. Rectangular iron shield boss attached to newly painted reconstruction of first-century AD Roman legionary semi-rectangular shield, by Legion Ireland. (Photograph by Marcus Vlpivs Nerva (Martin McAree) of Legion Ireland)

Plate 15. Reconstruction of first-century AD Roman legionary rectangular shield by Legion Ireland. (Photograph by Marcus Vlpivs Nerva (Martin McAree) of Legion Ireland)

Plate 16. Re-enactors of Legion Ireland demonstrating use of sword, spear and oval shield in simulated combat. (Photograph by Marcus Vlpivs Nerva (Martin McAree) of Legion Ireland)

Plate 17. A second picture of re-enactors of Legion Ireland demonstrating use of sword, spear and oval shield in simulated combat. (Photograph by Marcus Vlpivs Nerva (Martin McAree) of Legion Ireland)

Plate 18. Another picture of re-enactors of Legion Ireland demonstrating use of sword, spear and oval shield in simulated combat. (Photograph by Marcus Vlpivs Nerva (Martin McAree) of Legion Ireland)

Plate 19. Members of Deva VV (Chester Guard) carrying curved rectangular *scutum*, demonstrating throwing the *pilum*. (Photograph by permission of D. Flockton Photography)

Plate 20. *Testudo*: as displayed by members of Legion Ireland. (Photograph by Marcus Vlpivs Nerva (Martin McAree) of Legion Ireland)

Plate 21. A shield wall formation seen in a re-enactment by members of Deva VV (Chester Guard). (Photograph by permission of D. Flockton Photography)

UNIT	EMBLEM	BIRTHSIGN	FOUNDATION	RECRUITMENT
4TH MACEDONICA	Bull	Capricorn	Poss. Pompey Magnus	Orig. Spain or Italy
4TH FLAVIA FELIX	Lion (assoc. with Vespasian's favourite deity, Hercules)	Capricorn	AD 70	Dalmatia
4TH SCYTHICA	Bull	Capricorn	Late Republican. Legion of Pompey	Orig. Italy then Spain
5TH ALAUDAE (the Larks)	Elephants	Cancer	48 BC	Orig. Further Spain, & Alaudae auxiliaries in Transalpine Gaul
5TH MACEDONICA	Bull	n/k	Octavian. Pre-42 BC	Initially Spain, became Moesia
6TH FERRATA (Iron Clads)	Bull	Gemini (She-wolf & twins)	Republican. Orig. Pompey's 6th	Orig. Italy, later Spain
6TH VICTRIX	Bull	Gemini	Based on remains of Pompey's 6th	Italy & Spain
7TH CLAUDIA PIA FIDELIS	Bull	Leo (Lion)	*c.* 55 BC	Initially Spain. By Late 1st C. BC, Asia Minor.
7TH GEMINA	Bull	Gemini	AD 68	Eastern Spain
8TH AUGUSTA	Bull	Capricorn	Republican	Initially Italy, later Spain
9TH HISPANIA	Bull	Capricorn	*c.* 55BC	Spain
10TH FRETENSIS	Bull, Warship & Dolphin	Taurus	Disputed poss. 61 BC	Further Spain
11TH CLAUDIA	Neptune's trident & Thunderbolts	Gemini	58 BC	Cisalpine Gaul
12TH FULMINATA	Mar's Thunderbolts	Capricorn	58 BC	Cisalpine Gaul
13TH GEMINA	Lion	Capricorn	58 BC	Cisalpine Gaul
14TH GEMINA MARTIA VICTRIX	Eagle's wings & Thunderbolts	Capricorn	57 BC	Cisalpine Gaul

Table 4. Military units – shield emblems and birth signs (2) (from Dando-Collins, 2010).

UNIT	EMBLEM	BIRTHSIGN	FOUNDATION	RECRUITMENT
15TH APOLLINARIS	Originally Griffin. By 3rd C. Palm branch signifying Victory	Capricorn	54 BC	Cisalpine Gaul
15TH PRIMIGENEIA	Wheel of Fortune	Capricorn	AD 39	Cisalpine Gaul
16TH GALLICA	Boar	Capricorn	49 BC	Gaul
16TH FLAVIA FIRMA	Lion	n/k	AD 70	n/k
17TH, 18TH 19TH (Lost Legions of Varus)	Boar	Capricorn	49 BC	Italy
20TH VALERIA VICTRIX	Boar	Capricorn	49 BC	Initially Italy, later Syria
21ST RAPAX	Boar	Capricorn	49 BC	Initially Gaul, later Syria
22ND DEIOTARIANA	Eagle (used on coinage of King Deiotarius of Armenia Minor)	n/k	Formed from remnants of 2 legions of King D. Fought for Caesar, then Mark Antony.	Armenia Minor
22ND PRIMIGENEIA PIA FIDELIS	Eagle	Capricorn	AD 39	Probably in the East
30TH ULPIA	Neptune's trident, Dolphin & Thunderbolts	Capricorn	AD 103	n/k
PRAETORIAN GUARDS	Eagle & Thunderbolts		6th C. BC	
IMPERIAL SINGULARIAN HORSE (*EQUITUM SINGULARIUM AUGUSTI*)	Scorpion		AD 69	Hand-picked German cavalry

Table 5. Military units – shield emblems and birth signs (3) (from Dando-Collins, 2010).

		Width of cover	Length of cover	Board estimate
1	SMALL OVAL			
	Valkenburg 1	54 (60)	112 (123)	(45 x 104)
2	POINTED OVAL			
	Valkenburg 2	65 (71)	130 (143)	(60 x 120)
3	BROAD OVAL			
	Caerleon	76	–	(64)
	Bonn	–	–	–
	Oberaden	77	–	(64)
4	RECTANGULAR			
	Vindonissa Abb. 49	74	–	(64/7)
	Vindonissa Abb. 52	67 (70+)	–	(64/7)
	Doncaster	–	–	64 x 125
	Fayum	–	–	63.5 x 128
5	LARGER OVAL/ RECTANGULAR			
	Roomburgh 1	82		(75)
	Roomburgh 2	c. 82		(75)
	Vindonissa Abb. 59	82		(75)
6	**Dura Europos**	–	–	83 x 102
7	SMALL CIRCULAR			
	Castleton	60	60	48-50 diam.

Above sizes in cm, between brackets adding 10% shrinkage allowance for dry leather

Table 6. Comparison of shield cover/shield board sizes (from Van Driel-Murray, 1988; 1989; &1999).

VIII

CONSTRUCTION OF THE SHIELD BOARD

MATERIALS

No study of Roman military equipment would be complete while restricting discussion to the armour alone. Armour is the end product of numerous interrelated industries, from the production of the raw materials, their transportation, primary and secondary processing, finally down to the armour's construction. This by necessity draws into the discussion at every stage the questions of 'what', 'how', 'where', and 'by whom'.

Before considering the processes of construction of shields and shield furniture, it would be helpful first to step back and take an overview of the materials used (wood, leather and metal), the primary and secondary processes involved in production, and what signs (if any) would remain of these in the archaeological record (see appendices III, IV and V: WOOD; LEATHER; METAL). This could help in the identification of potential locations, both for sourcing of raw materials, any primary and secondary processing, and finally for manufacture, storage and distribution of the end product. This would then by necessity encompass a wide range of different materials, not all of which would be naturally occurring in all parts of the Empire, suggesting the favouring of some locations over others for specific armour/equipment types and also for pre-established, well-developed supply/distribution infrastructure.

For example, for metallic parts of equipment, raw materials could not be sourced locally everywhere, some locations being closer to deposits of fuels and ores (whether iron, copper, zinc, tin, etc.), or to locations known to be established for the recycling of scrap material, or with a tradition for skilled craftsmen/armourers. Other equipment may contain material that can be sourced more readily, for example wood for shields, or leather for belts, tents, strapping for plate armour, etc. However, even with these obviously more common materials, preference may have existed for sourcing from some locations over others, for example with hides shipped in over considerable distances, or wood types not native to an area being chosen for their properties of strength, density or elasticity, in preference over timber more readily available locally.

MATERIAL PROCESSING (SUMMARY)

It is apparent that the three material types involved in the production of shields, and the other main components of the Roman military assemblage (wood, leather and metal), while all being important components, will each leave different levels of archaeological evidence of their use.

Wood residue as a by-product of production, being a natural material, is difficult to differentiate in the archaeological record from that naturally occurring, or produced as a residue of construction work, Roman or otherwise. Evidence of its use may present therefore as tools (often similar to those used for other processes). Alternately, it may be possible to locate large, relatively featureless workshops, which may have been used for large-scale production of shields, particularly at potential factory sites which may have served to equip new legions in bulk. To date, no such high-capacity production sites for shields have been identified, although it is possible that future work, perhaps on pollen analysis, may be able to locate large areas of woodland clearance, datable to contexts which correlate to periods of intensive production.

Leather processing sites can be similarly difficult to locate, as they may be intentionally sited at distance from areas of habitation due to the unpleasant nature of the materials involved. However, where found, common features of tanneries would consist of an adequate water supply, water tanks and vats for soaking hides, associated possibly with external areas containing multiple postholes from stretching and drying racks. Secondary evidence of leather processing, in the form of offcuts, horns or horn cores, may be suggestive of manufacturing sites or, if found with quantities of worn fragments, may derive from small workshops involved in low-level production and repair work. Most manufacturing sites appear to be set back behind frontier lines, with the suggestion that hides were temporarily dressed with salt at point of slaughter, passed back to processing and manufacturing sites in more secure locations, and finished goods then sent forward to the frontier installations.

Uniformity of design in the earlier Imperial period (suggestive of military-run manufacturing sites) later relaxes into a greater variety of more 'civilian' styles of footwear (suggesting a greater level of non-military, private enterprise at play). In view of the volume of shoes which would have been required each year to supply the military, along with their more innocuous nature compared to weaponry and armour, it is not surprising that these items in more peaceful times would have been those primarily farmed out to indigenous local suppliers. The supplies of metal and weaponry/armour production, however, would have been strictly monitored and overseen, if not directly manufactured by the military.

The identity of manufacturers of metallic armour and weaponry changes its focus periodically, reflecting the changing nature of the Roman army and significant events in the history of its people. In the early Republican period and continuing down to the time of the Punic Wars, each man provided his own equipment, commissioned from local craftsmen working in small-scale workshops. Furthermore, armour at that time was predominantly fashioned in bronze, with lesser use of iron (in the form of mail and some later-period helmets of Gallic influence).

At times of extreme measures, when manpower was at its lowest, with the need to recruit large numbers of men into new legions, as with Fabius Maximus' recruitment of 8,000 slaves after the defeat at Cannae (Cottrell, 1992, 153), or Marius' raising armies for his African campaign from men of the poorer *Capite censi* (Scullard, 1970, 52), armour and weapons would have been mass produced at state expense, probably in state-controlled 'factories'. The locations of these are not known but design features suggest the use of Gallic craftsmen. Similarly, Julius Caesar also appears to have made use of Gallic craftsmen to re-equip his men during his Gallic Wars, the level of production again suggesting the existence of military-controlled 'factories'.

By the Imperial period, the nature of military supply had changed completely, with men no longer supplying their own equipment, the army providing each recruit with his basic kit on enlistment, paid for by deductions from pay. This 'basic' kit would probably have been commissioned 'in bulk' by the military from its own large-scale manufacturing sites, such as Templeborough. The individual soldier would, however, augment this basic kit by purchasing better-quality 'bespoke' equipment, probably produced by civilian craftsmen in small-scale workshops, using standard designs but including their own distinctive cultural or personal features.

Four different types of site can therefore reasonably expect to be located. These would be primary processing sites, where metal was produced from the ore; large-scale manufacturing sites; small-scale civilian workshops (in purely civilian settlements, or in civilian *vici* outside forts and military installations); or in military workshops, or *fabricae* (inside the forts, or in defended military annexes attached to a forts defences). These sites would be identifiable by the quantity and types of furnaces and hearths present, along with the presence of water supplies, water-storage tanks, residual traces of ores, fuel, slag, hammer scale, metal fragments and specific building types.

REPRODUCTION OF THE SHIELD BOARD

From the literary and archaeological evidence, a basic constructional methodology can be seen in the production of both plank- and ply-built shield boards. Polybius reports the best types of wood to use, indicating that Roman craftsmen were only too aware of the properties best suited to the intended purpose – which dense, hard woods were best for strength and rigidity, and which were softer, more flexible, easier to shape and to close up around a penetrating weapon. It is clear, however, from the range of wood types identified at Doncaster, the Fayum and Dura, that although some ideal wood types were preferred, in practice craftsmen would select the most suitable materials from those most readily and locally available. Whatever the wood type chosen, its properties would be further influenced by the way the planks or strips were cut from the trunk.

Roman methods of processing a tree trunk differ from the modern. In a modern sawmill, the trunk is rough sawn into planks, dividing its length by cutting tangentially, before drying and seasoning. However, in antiquity this was not the practice. The trunk or butt would be divided lengthways into quarters by splitting (by forcing wooden wedges along its length using a heavy mallet), then stacked and allowed to dry before cutting into planks radially, at right angles to the grain lines (Zant, pers. comm., October 2005). The properties of the wood itself will further be affected by the part of the trunk from which the plank is cut.

Living wood is composed of cellulose-walled cells, containing living matter in the centre. As it grows, the tree lays down layers of these cells at its outer edge, under the bark. This sapwood is moist, sappy and more flexible than the inner part of the trunk. In the older cells, towards the centre, the cellulose is replaced with a material called lignin, and the central living part dies. This heartwood is drier, harder, but less flexible than the sapwood. The Roman craftsman would have been well aware of these properties, and would have no doubt selected accordingly – heartwood being more suitable, by its strength and rigidity, for middle ply layers, but the more flexible sapwood being preferable for outer ply, where flexibility was required.

The planks would then have been roughly trimmed, with either an axe or an adze, both tools leaving similar tool marks or 'facets', as identified by James on the Dura examples (James, 2004, 160). Further fine smoothing of the surface, or thinning down the thickness of planks used towards the outer edges of the shield, could have been achieved with a plane, or more probably a drawknife, again producing the tell-tale 'chatter' marks found by James on the Dura shields (James, 2004, 160).

Any reconstruction would, as acknowledged by Buckland (1978, 257), involve a degree of compromise between the ideal situation of all materials produced following fully authentic traditions, and those equivalent materials currently available through modern methods. Therefore, in his reconstruction of the Doncaster shield, Buckland chose to use modern three-layer plywood (Fig. 33). If one accepts his interpretation of the board as flat, this would not have compromised too greatly the final reconstruction visually, although, as he accepted, the handling properties of weight and response to penetration weapons would have differed (Buckland, 1978, 259). However, to produce a curved board from this material does pose a different range of problems than would be encountered using the correct Roman materials. It was suggested by curatorial staff at Doncaster Museum, when the original reconstruction was first displayed, that initial attempts had been made to produce a curved board, but that this had been pulled back to straight by the stretched leather surfacing (J. R. Travis, pers. comm., 2000).

Some reconstructions of curved boards by re-enactment groups have been achieved using modern ply board, however, by use of either of two methods: by pressing into a concave mould, or by flexing over a convex former (L. Morgan, pers. comm., July 2005). By using two layers of thin ply, with a layer of strong, freshly applied contact adhesive in between, placed into the mould/over the former and secured until the glue sets, the curved shape will be retained. By the subsequent addition of a grid of thin cross-braces, as seen on the Dura examples, the shield board is given a greater rigidity, while further mitigating against loss of curvature (Fig. 41).

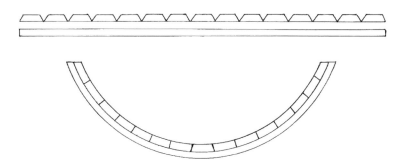

Fig 64. Constructing the plyboard shield board: structure of ply. (Artwork by J. R. Travis)

However, in antiquity, the Roman shield maker would, in all probability, not have worked with preformed ply. The only reason that the modern ply shield retains its shape is because two layers are used, bent to curvature before the glue cures, the outer layer being slightly larger than the inner, describing a greater circumference. For the Roman shield maker to produce a shield in this way, from preformed ply, he would have needed still to place two separate sheets into the mould/over the former. This then would have produced a board with a minimum of four layers (assuming ply formed from the minimum of just two layers). There is however, no evidence archaeologically, from Doncaster, the Fayum or Dura, of any more than three layers being used (possibly on one example from Dura and several from Masada, only two layers being present), so these methods must be discounted. If a ply board were to be constructed from two or more layers of alternately oriented strips/planks, laid down in the flat, and then attempts were to be made to put this into a mould or over the former, with the glue still wet, it would disintegrate. It can therefore be assumed that the ply board bases of the Roman shields were constructed using two possible methods: i) within a concave mould; or ii) over a convex former (Fig. 64).

It was initially considered that the most probable method of construction would have been over a convex former, similar to the most popular method for construction of modern reconstructions. To test this method through reconstruction, a convex former was constructed, a small horizontal ledge along each side edge, with a small lip forming a slot to accommodate the strip ends. The middle vertical layer was then built up around the first horizontal layer, the edges of the planks being angled to allow close fitting as part of the curved structure. It was not possible, using this method, for the two outer vertical planks to be inserted into the edge slot, so the inner horizontal layer at this stage extended beyond the desired end width. Further difficulties were encountered in the application of the outer horizontal layer, each strip having to be glued and clamped individually to prevent separation (suggesting that the use of pins or wooden pegs, as seen at Masada, may have been commonplace). Once all pieces were applied and the glue fully cured, the board was removed from the former, the edges trimmed, and a grid of reinforcing strips applied to the rear surface (although care was necessary to ensure that the board did not unfold before the underlying grid could be applied). Finally, the hand hole was cut in the centre and the shaped handle attached to the rear, which also acted as a further horizontal cross-strengthening. Although this method did produce a useable board, the process was very slow and would not have been an efficient method for any kind of mass production.

Attempts to produce a board using a concave mould were almost equally unsuccessful. Although it was possible to construct a mould and to secure the first layer of flexed horizontal strips within it, the second vertical layer posed difficulties, in the application of a uniform pressure to hold against the first layer while the glue cured. To assemble all three layers before the glue solidified was still more problematic, but did allow then for a central upright bar-press to be wound in, forcing the layers together, with side clamps to prevent edge separation. Once the glue was cured, the outer clamps were removed and a grid of reinforcing strips applied to the rear surface as in the Dura example. Although it did prove possible to produce a board using this method, it was nevertheless fraught with difficulties. In order for the board to be assembled sufficiently quickly, within the drying time of the adhesive, all component strips had to be accurately cut to size in advance, and laid out in order for rapid application. In conclusion, this was a most impractical method of manufacture, which would have had an unacceptably high failure rate and subsequent material wastage, the effect of which would have increased in warmer climates, perhaps to total unfeasibility.

From experiments with the two methods above, a third, hybrid method was developed. An open, vertical framework was constructed, with two flat horizontal boards separated by a series of vertical struts. A curved slot was cut through each of the flat boards, the shape and dimension of the shield board profile. The first stage of the shield board's construction was then completed flat on the ground, without the use of the frame. The horizontal planks/slats of the outer layer were laid out on the ground. The vertical planks, their side edges sloped as in the previous method, were then glued edge to edge across these horizontals, their widest surface downwards, forming a V-shaped groove between each. Once the adhesive was cured, a two-ply board was produced, highly flexible in one direction only. The V-shaped grooves were then coated with adhesive, the board curved with the vertical planks in the inside position, closing up the grooves, and the whole piece slotted into the frame through the two horizontal curved slots. The third layer of horizontal planks/slats was then glued in place on the inner surface, omitting the upper and lower edges, retained by clamping at the edges. Once the adhesive was again cured, the underlying grid could be fixed as before, ensuring that there would be no loss of curvature. Once removed from the frame the final upper and lower rear pieces were also applied (Figs 64–66).

This last method again produced a satisfactory end result, but in a shorter time, with more efficient use of the frame. However, there is no archaeological evidence for any method of construction, so any conclusions are entirely conjectural, and it is entirely possible that some other, unconsidered method was used. A further consideration, which it was not possible to factor into the experiment, was that the Roman shield maker was a skilled craftsman, as opposed to the enthusiastic amateur of the modern day. Difficulties encountered within these trials may have been due more to a lack of skill and experience than to a weakness in the methodology. Other problems associated with drying/setting times of adhesives may also have been a result of the use of modern equivalent materials (contact adhesive), the Roman glues perhaps again exhibiting different properties.

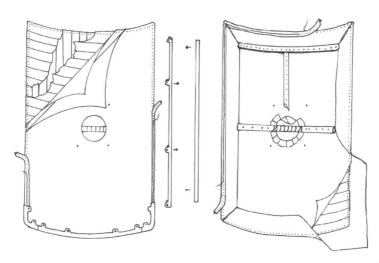

Fig 65. Constructing the plyboard shield board: (a) front view; (b) rear view. (Artwork by J. R. Travis)

Roman glues would have been organic, although possibly not vegetable-based as these may not have offered sufficient strength – more likely they were fish- or animal-based. Animal-based adhesives have been used throughout history, in concentrated form as glue, and in more dilute form as a size, to seal surfaces prior to application of painted surfaces. This is traditionally made by boiling up rabbit skins, animal hooves and similar inedible animal by-products. It would not only provide a strong bond, holding together the layers of ply, but would serve to fuse the board to a surface fabric or leather coating, sealing the surface prior to painting. Being of animal origin, on combustion, this type of glue could appear visually and chemically similar to the 'bubbly char' reportedly found on the surface of the Doncaster shield board (Biek, 1978, 269).

As the planks/slats of wood used to produce the plywood shield board of the reconstruction were very thin (less than 3 mm), they flexed easily. However, the broad, plank-built oval shields found at Dura, being made from much thicker, less flexible planks, would have required some additional treatment to shape the boards, probably steaming to temporarily soften the fabric of the wood. If flexed over the former immediately after steaming, and secured until cooled, the wooden planks would retain their shape. In other Roman plank-built constructions (boxes and boats), small slivers of wood are inserted into slots cut into the joining edges, helping to strengthen the bond. However, there is no evidence that this method of construction was employed on the Dura shields, and it is entirely possible that the individual planks were merely glued to their neighbours. If this were the case, it would suggest that the adhesive properties of the Roman animal-based glue were far superior to the modern equivalent.

PREPARATION OF THE SHIELD BOARD

Although Buckland reports that the Doncaster shield was probably leather covered, based on his visual interpretation of the remains, the chemical analysis neither proved nor discounted this interpretation. The shields at Dura, however, were found in some cases to be leather coated, but in other cases this was not the case, with a woven fabric glued to the surface, covered with a gesso-like substance before painting. For the

Fig 66. Making the shield board: (above) laying out the vertical strips and applying the inner ply of horizontal strips; (below) first two layers of ply assembled and ready to shape in convex press. (Photograph by J. R. Travis)

Fig 67. The reconstructed legionary shield covered and primed, boss attached, ready for painting. (Photograph by J. R. Travis)

sake of economy, the latter method was chosen on the experimental board, the board being covered with a thick coating of animal-based size, then a fine linen fabric stretched over the shield (Fig. 65), with a further heavy top coating of size to finish. Once dried, a thick coating of gesso was applied to seal the surface, evening out imperfections, prior to painting with earth-based pigments in a tempura medium (Figs 66–68; for a more in depth discussion on the manufacture of size, gesso, paint, pigments and dyes, see Appendix 6). This treatment of the board surface not only reproduces the archaeological finds from Dura, but also reflects methods and traditions carried down throughout the Roman, medieval and Renaissance periods for the production of boards for panel paintings.

Finally, the reconstruction was completed by the attachment of the shield boss and edging strip (using lengths of wide, flat brass strips, folded to form a U-shaped channel, with protruding riveting lugs at appropriate fixing positions), the former attached using large, flat-headed copper rivets, peened on the inside surface over copper washers, and the latter using smaller, round-headed brass rivets, peened onto the rear surface of the strip.

Fig 68. The reconstructed auxiliary shield: (a) covered and primed with gesso; (b) rear view. (Photograph by J. R. Travis)

IX

DESTRUCTIVE FIELD TESTING OF THE RECONSTRUCTED BOARD

The reconstructed shield described in the previous chapter was then subjected to field testing, in order to consider the practicalities of its use, through assaults with sword, *pilum* and arrow strikes.

LIMITATIONS OF 'FIELD TESTING' REPRODUCTION EQUIPMENT

Any attempts at 'field testing' reproduction equipment would need primarily to recognise a multitude of possible limitations, based on a wide range of variables within the test equipment. Within any tests, the accuracy of any results is proportional to the sample size used. However, regardless of sample size, the value of any results produced would be reduced to a far greater extent by any divergence from the accuracy of the test conditions – factors such as materials and equipment used, weather conditions and abilities of the tester. Ideally therefore, the shield to be tested and the equipment used for the test 'assaults' should be reproduced from materials matched as closely as possible to the originals. The 'tests' should then be performed by a tester of comparable competence to the original users, under a range of possible weather and geographic conditions, and repeated to produce a significantly large bank of test 'data'. It therefore has to be recognised that any of the above factors may be manipulated, intentionally or unintentionally, in order to skew test results to favour an anticipated predicted outcome. With these caveats in mind, we may consider each of the main limitations of the 'test':

• Accuracy of the shield materials
• Weather and geographic conditions of the tests
• Ability of the 'tester' and accuracy of the 'test' weaponry: the bow; the arrows

Accuracy of the Shield Materials
Any test of the effectiveness of an item of armour against a weapon strike would be affected by the quality of both the weapon and the armour itself, in constructional details and materials used. Some 'tests' have been carried out by members of the re-enactment community on reproduction shields, although as re-enactor shields are made from preformed modern plywood, and not from authentic materials, the results may not be conclusive. While the highest level of accuracy can be maintained in constructional features through close consideration of existing artefacts, it is never possible to totally reproduce the materials used – leather will have been produced from modern tanneries, with hides from animals culled at different life stages; metal will have been smelted from different ore streams, with modern-day background pollution levels; wood will be modern ply, of wood seasoned using modern methods, and not using the types of wood known to have been used in antiquity (which, being of different hardness, would react differently to an assault).

Provided the nearest equivalent materials are sourced, and that constructional methods are as close as possible to the originals, the performance of 'test' items may still be assessed objectively. As described in earlier chapters, all leather used in the reconstructions was produced using vegetable-based tanning processes, and the metal boss was produced using mild-steel plate, heated and hammer-forged to shape.

Weather and Geographic Conditions of the Tests

As the Roman army operated across a widely varying geographic area, it would no doubt have encountered the full range of weather and geographic conditions, and would have had to contend with all of these. However, it is recognised that heat and humidity will have effects upon the performance of archery equipment: the elasticity of bow limbs is different when warm (with archers, where possible, warming the limbs of the bow before use); the bow string will similarly be susceptible to extremes of temperature and moisture (hence the term 'keep it under your hat' deriving from where archers traditionally kept spare bow strings warm and dry); and arrow shafts and fletchings will also perform differently in the cold and wet. The distances achieved by an archer, using the same bow and the same arrows, will also be greater when shooting downhill rather than uphill, so results in hilly regions will be different to those on a flat shoot.

   For the purpose of the shield 'tests' therefore, test shoots were performed at different times of the year, under a range of climatic conditions, with targets at a range of distances, some uphill, some downhill and others as flat shoots, to produce a range of 'average' results for a selection of different arrowhead types.

Ability of the 'Tester' and the Accuracy of the 'Test' Weaponry: The Bows; the Arrows

We can assume that, for the most part, the effect of the tester's abilities on the test outcomes will be minimal, provided that the tester is able to hit the target at the required distance. In this respect, the 'tester' (J. R. Travis) is a historical re-enactor, with over ten years of experience in the use of medieval longbow, in addition to representations of both a Roman legionary and auxiliary archer, using a range of bows, including Scythian-style recurved bows. However, due to health and safety constraints, the bows used in modern re-enactment are deliberately restricted to 30–40 lb strength, whereas the exact poundage of Roman period bows is unknown, with speculation that some may have been more powerful (it is known that the medieval longbows were indeed of considerably higher poundage).

   Without intending a full discussion of the function of archers (*sagittarii*) or archery units (*sagittariorum*) within the Roman military, which is too extensive a subject to be discussed in depth in this particular publication, a brief description of archery equipment available will be necessary, in order to consider testing the effects of archery projectiles (arrows) on the shield. This will require a discussion of not only the bow types used, but also the arrows (shaft materials, arrow head type, weight and hardness).

BOWS

There are two types of bow known to have been in use during the Roman period: self bows (made from a single piece of wood) and composite bows (wooden core, sandwiched between layers of horn on the inside and sinew on the back), first invented by the nomadic peoples of the Asiatic steppes. The sinew and horn stores energy when drawn, delivered to the arrow on release, so that comparable power can be produced by a much shorter bow than an equivalent self bow. Composite bows are generally recurved, storing further energy and producing a greater final draw weight. The ends of the bows would be stiffened by attachment of bone or antler laths, and these, being more resistant to decay than the wood or sinew components, are often the only parts found in archaeological contexts.

   Vegetius (*Epitoma rei militaris*, 1.15) proposed that legionary recruits were trained with 'wooden bows' (*arcubus ligneis*), which may have been self bows, possibly using the European yew (*Taxus baccata*), like the later northern European 'longbow' (although it is not known whether these matched the medieval longbow in strength or performance properties). Vegetius also cited Scipio Africanus, who, describing Numantine opponents, 'did not consider himself superior, unless he had mixed selected archers in all centuries' (*ibid.*).

   The bows in use by auxiliary archers within the Roman army may, however, have been constructed using their own ethnic traditions (possibly composite bows, as was known to have been used by many of the nomadic Central Asian horse archers, including the Scythian, Sakas, Sarmatians and Parthians), with bow laths being frequent finds in many Roman military contexts (Coulston, 1985, 220–366).

While bow laths are found from a number of Roman context sites, including Carnuntum (von Groller, 1901b), complete bows are rare, their component parts being organic and not readily lending themselves to preservation in most soil conditions. A couple of notable exceptions are the 'Yrzi' and the 'Qum-Darya' bows. The Yrzi bow, from a necropolis context at Baghouz, on the Euphrates, has been identified as a Parthian weapon, dated to between the first century BC and the third century AD, based on associated finds. Unfortunately, although almost complete, one arm was missing. Various reconstructions have been attempted, but its attribution as a 'horse archer' bow (for 'warrior' or 'hunting use), of 'Eastern Parthian' or 'Western Parthian' origin, hinges on whether it should be reconstructed with a symmetrical or asymmetrical shape (with limbs of different lengths). This has a direct bearing on its estimated draw weight, with Coulston (1985, 240) proposing 60–70 lb draw weight symmetrical reconstruction, while James (2004, 19) proposed a greater strength of 80 lb, with an asymmetrical shape better suited to horse archer use.

The 'Qum-Darya' bow was found in a mass grave associated with a Chinese Turkestan frontier post (the Tibetan fortress of Mazar-tagh at Khotan-darya, in the Lop Nor region of Xinjiang). Like the Yrzi example, it was a recurved composite bow, similarly dated from first century BC to third century AD, although styled in 'Hunnic' tradition, with straighter ears, and using a greater number of ear and grip laths – up to seven, compared to only four on the Yrzi bow, although this may be due to the latter being considered of poorer-quality manufacture (Bergman, 1939, 121–124). Again, unfortunately, the bow was poorly preserved, with only a quarter remaining, so estimations of its poundage would be speculative, save for the knowledge that it may be comparable to the similar Yrzi example above.

ARROWHEADS

Literary sources suggest the efficiency of arrowheads in use during the Republican period in armour penetration, for example in both Plutarch's and Cassius Dio's reports of Crassus' experience of Parthian arrows at Carrhae:

> when Crassus ordered his light-armed troops to make a charge, they did not advance far, but encountering a multitude of arrows, abandoned their undertaking and ran back for shelter among the men-at-arms, among whom they caused the beginning of disorder and fear, for these now saw the velocity and force of the arrows, which fractured armour, and tore their way through every covering alike, whether hard or soft. (Plutarch, *Crassus*, 24)

> The missiles falling thick upon them from all sides at once struck down many by a mortal blow, rendered many useless for battle, and caused distress to all. They flew into their eyes and pierced their hands and all the other parts of their body and, penetrating their armour, deprived them of their protection and compelled them to expose themselves to each new missile. (Cassius Dio, XL.22)

However, in consideration of the above reports, it should be remembered that, at this time, not all combatants would be wearing the same level of armour as in later periods (mail was not universally used, some still using the less efficient simple breastplates, offering only minimal protection, and *segmentata* was not yet developed for heavy infantry), the second quotation above, from Cassius Dio, describing a range of soft-tissue injuries to unprotected parts of the body. Further, we are unable to determine the accuracy of these reports in the ratio of arrow injuries sustained compared to those received in hand-to-hand combat or from cavalry assault, which would also have featured in this particular event. In addition, the sources do not relate the types or range of arrowheads in use, in that archers may have made use of untipped or lightweight headed arrows to wear down an opponent in an initial 'arrow storm', before using heavier weighted specialist arrowheads for specific targeting of armoured opponents at closer range.

The most common form of Roman arrowhead was trilobate (mostly made of iron and tanged, although with some earlier forms in copper alloy and socketed), with average blade length of around 2.5 to 6.5 cm, not including tang, although bilobate and leaf-shaped forms are also known. Although it is not proven which shapes, if any, were intended specifically for war or hunting, the trilobate forms were probably well suited to both purposes, being more aerodynamic and more accurate. They also caused greater injury and would have been more difficult to remove from an unarmoured target. However, there were other 'eastern' forms of arrowhead in use, in addition to some not dissimilar to London Museum types 7, 8 and 10 medieval-period 'bodkins', and some other heavy piled forms, which appear in later-period contexts, any of which could have been used for armour-piercing purposes.

A comprehensive discussion of Roman archery equipment can be found in James's report on the arms and armour at Dura Europos (2004, 191–208), where remains of both bows and arrows were found. Although no complete arrows were found, the site produced fragments of all component parts of numerous arrows, some iron or copper alloy tipped, others wooden tipped, and with one example still retaining its fletchings. James discounted the large number of copper alloy socketed arrowheads found as probably being 'residual' deposits, predating the Roman occupation by several centuries (2004, 195). He similarly discounted the two flat-type iron arrowheads as being later-period 'surface finds' (2004, 195), identifying the remaining tanged iron arrowheads with three barbed, triangular blades as being from the Roman period, although probably in use by both attacking and defending sides.

James (2004, 195) also proposed that many of the arrows found at Dura Europos were more simply constructed, with wooden tips rather than metal arrowheads, as these had been 'very common in the ancient world', either for training purposes (citing Arrian's description of wooden training javelins, in *Ars tactica* 34.8, 40.1), or for general use, where he considered them 'adequate' against unarmoured men. In support of this suggestion, he noted that a number of these wooden-tipped arrows exhibit signs of decoration, so should not be considered as 'emergency' replacements, which had been pressed into service before time was found to finish them off by the addition of metal arrowheads.

Although discussing medieval-period equipment, *De Re Metallica: The Uses of Metal in the Middle Ages* (whose title 'about metallic things' is taken from the sixteenth-century treatise by Georgius Agricola), is a collection of twenty-three papers, including that by David Starley (ed. Robert Bork, 2005). Starley, using metallography based on a very small sample of four type 16 broadheads and two bodkin heads, proposed that late medieval arrow makers developed 'high-tech' steel-enhanced broadhead arrows in order to counter the advance from mail to plate armour. He cited Jessop's typology of medieval arrowheads (1996, 192–205) to suggest that bodkin heads were intended for use against mail.

However, it is equally possible, based on Halpin's study of medieval arrowheads from Ireland (Halpin, 1997), that heavier weighted bodkins may have been introduced in the medieval period for the purpose of piercing plate armour, and that these may not necessarily have been hardened, with weight and diameter of projectile being more significant than hardness. Test shooting of type 8 bodkins and type 16 broadheads of the same weight, using the same bow, has found that the former, being more aerodynamically shaped, exceeded the latter in distance. In terms of piercing plate armour, however, the heavier type 7 bodkins performed best, but would have only been intended for short-range shooting.

Although hardened arrowheads may be preferable for piercing armour, it is not proven that all non-hardened arrowheads were not so intended. Further, as the sample range of tested arrowheads is so small, it is not possible to say with any certainty that the medieval broadheads were the first and only arrowheads to be hardened. Certainly the technology to produce hardened heads was not unknown to Roman blacksmiths, so, until a comprehensive metallurgical study is carried out of Roman-period artefacts, this cannot be proven either way. However, it is equally possible that, when producing large numbers of arrowheads at short notice, time constraints may have necessitated the majority of points being unhardened in both Roman and medieval periods, with the additional possibility of some untipped arrows in combat use.

'FIELD TESTING' ARCHERY EQUIPMENT USED

For the purposes of testing the reconstructed equipment above, therefore, both tipped and untipped arrows were used, test shooting from distances of 30 feet and 60 feet from the target (approx. 10 m and 20 m). For the 'test' archery assaults, a replica first-century AD Scythian-style recurved bow of 35 lbs draw weight was used, with arrows spined to 30–35 lbs (this being close to that which may have been available to horse archers of the Roman period, although possibly not as powerful as the Yrzi example above, but limited by modern re-enactment health and safety regulations). The fletchings observed on the arrows from Dura Europos were approximately 6 inches long (15.5 cm), but were of a very low, rounded profile (only approximately 1 cm in height), which would have projected a fast, straight course, with maximum penetration. For test-shooting purposes, arrows were used with a range of possible arrowhead types, with similar fletching length to those from Dura Europos, although with fletchings set further from the nock, for conventional 'standard' release, rather than 'thumb ring' release (as had been the case with the Dura examples).

When an arrow is shot from a bow, the stored energy in the bow transfers into kinetic energy in the arrow and the bow. This can be shown using the following mathematical calculations (Rees, 1995), where:

F = Force required to draw bow
x = Distance bow is drawn
e = Estimated efficiency of bow = 0.9
M = Mass of the bow
k = Factor representing the kinetic energy of the bow (0.03 to 0.07, depending on bow)
m = Mass of arrow
v = Speed of arrow
g = Acceleration due to gravity
d = Maximum distance achievable

Stored energy in bow = $eFx/2$ (using Hooke's Law)
Kinetic energy in arrow = $(mv2)/2$
Kinetic energy of the bow = $(k\ Mv2)/2$

Thus:
Kinetic energy in arrow + Kinetic energy of the bow = Stored energy in bow

$(mv2)/2 + (k\ Mv2)/2 = eFx/2$
$(mv2) + (k\ Mv2) = eFx$
$v2\ (m + (k\ M)) = eFx$
$v2 = (eFx/m + (k\ M))$
$v = (eFx\ /\ (m + kM))-2$
$d = v2/g$

The maximum range (d) that can be achieved (by aiming at 45° to the horizontal) depends on the initial speed of the arrow (v), and the acceleration due to gravity (g). The true maximum distances achieved will however be less, due to the drag caused by air resistance on the arrow, although for purposes of calculating equivalent distances between different poundage bows, the discrepancy can be ignored.

The calculations above were derived by Rees (1995) to gauge comparative capabilities of English longbows, using an estimated 154 lb strength. They can, however, be used to extrapolate values for other weights of bows. The Yrzi bow is described as being in the range of 60–70 lb (Coulston, 1985) and 80 lb (James, 2004); from the above calculations it can be seen that for the test bow of 35 lb to produce an equivalent impact, its distances to the target would need to be decreased by 50 per cent. The results then shown in Table 7 would represent those given by an equivalent bow to that from Yrzi at double the distances indicated.

'FIELD TEST' RESULTS

A breakdown of average test results using a range of arrowheads against a wooden shield is provided in Table 8 below. For example, it was found that the metal-tipped ('bodkin'-headed) arrows, flat shot at a distance of 9.23 m (30 feet), were able to easily penetrate a 10 mm-thick, three-ply wooden shield, the arrow totally clearing the shield, although penetrating the shield by a depth of only 24 cm from a distance of 18.28 m (60 feet). At the same distances, even the wooden-tipped arrows penetrated the same shield by 45 cm and 15 cm respectively, although some arrows striking the curve of the shield at a tangent were deflected to the side. Clearly, the main function of the shield would have been in its use as a weapon, and in its ability to deflect weapon strikes in hand-to-hand combat, or to collect untipped arrows from an 'arrow storm', the energy of these arrows being almost spent. They would have offered little protection against a close, flat-shot, direct arrow strike, other than as a first line of defence, slowing the arrow before it struck the mailed, padded body behind it.

Clearly some arrowheads (particularly bodkin types) were more efficient at armour penetration, others were better at causing soft-tissue injuries (broadhead and barbed types), whereas the lightweight, untipped arrows were best suited to arrow storms, where they may cause *ad hoc* injuries to exposed areas of flesh (face, arms, legs, hands, feet, etc.) or may render a shield an unwieldy encumbrance. The untipped arrows were, however, virtually useless against any level of body protection. They would also have been of little use when shooting from higher-poundage bows, as they would in all likelihood disintegrate under the stresses of release.

The existence of these different arrow types would therefore suggest that both lightweight and heavyweight bows were in use, with arrow-storm assaults being used to cause the opposition to raise their shields, exposing unprotected or less armoured body parts (such as underarms) to more direct shots by more substantial arrow types. Alternately, longer-distance arrow-storm assaults by higher-poundage bows would similarly cause the enemy to raise their shields, exposing the less armoured areas to lightweight arrows from waves of horse archers with smaller, lighter bows. It would then follow that the use of this variety of arrow types and bow weights suggests that padding was probably in use to some level, to augment the efficiency of the armour (which is easily compromised when used without padding).

TESTING RECONSTRUCTED SHIELD AGAINST *PILUM* ASSAULT

Although with its primary focus being on the performance of the various known types of Roman *pilum*, in 1998, Connolly (2000, 43–46) conducted experiments testing penetration of six different *pilum*-head types in potential use against Roman-type shields of ply construction, comparing their relative penetration and likelihood of bending on impact. Quite separately, an impromptu *pilum-versus*-shield 'test' was similarly conducted during a training weekend by the Chester Guard re-enactment group, '*Deva VV*', at an event at Bishop's Castle in 2002, and, although light-hearted in nature, its results (in some ways different, but not at variance, to those of Connolly) have a relevance to this study in the discussion of shield performance.

In his experiment, Connolly tested six different types of reconstructed *pila* heads, three with barbed heads (one of 'Talamonaccio' type, and two of the slightly longer 'Smihel' types: types 2 and 3), two with pyramidal heads (the short *hasta velitaris* and the longer 'Renieblas' types), and the considerably longer socketed type. The *Deva VV* test, in contrast, made use of only one type of weapon, with a pyramidal head similar to the Renieblas type, but with a slightly shorter shaft (approximately 2/3 of its length, similar in length to Connolly's barbed Smihel type 3).

The *Deva VV* test of '*pila versus* shield' was conducted during a training session when soldiers were 'dared' to throw their *pila* and hit a shield, with the result that the shield was almost totally destroyed. The test used long, metal-shafted *pila*, equivalent to Connolly's Renieblas type and long, socketed type (although with shorter shafts, about two-thirds of their length). During the test, three *pila* went right through the shield, some up to the wooden stop; eleven of the *pila* stuck in the front of the shield; and some *pila* missed the target (fell short or went over the top), but there were no 'bounce offs'. It was also found that, despite what had been anticipated, none of the *pila* had been bent, either through impact with the shield or through hitting the ground. It should be considered, however, that the *pila* may have been over-engineered, causing them to be stronger than the originals. However, as the shield had also possibly been over-engineered, using modern ply of woods with different properties, and possibly thicker than it should have been, one would have expected more *pila* to have bent on impact.

In his test, Connolly (2000, 44–45) reported that the Renieblas type of head also passed straight through the target with no bending, but that they sometimes bent when they hit the ground. It is possible, however, that the *Deva* shafts were just lucky, or were more resistant to bending with their slightly shorter shafts, or may have been made from slightly stronger metal. In contrast, however, Connolly reported that the shorter Talamonaccio type failed to penetrate (but this had the larger, triangular-barbed head), and that the small *hasta velitaris* always bent without piercing the target (*ibid.*).

Connolly also used two of the Smihel-type *pila* (types 2 and 3), also with barbed heads (in common with the Talamonaccio type), one with a larger head but shorter shaft, and one with a smaller head but longer shaft, but he did not report their success in the throwing trials. However, in his test of piercing capability, he reported that these were difficult to test due to their bending and breaking at the tip, but that penetration was less efficient than with pyramidal types (such as the Renieblas and long, socketed types), suggesting that barbed heads in general were inferior against shields and would bounce off without much penetration (Connolly, 2000, 44–45).

His test throws used an experienced javelin thrower (against 11 mm-thick, flat, three-ply wooden board, rather than a reconstructed semi-circular shield), achieving distances of 54.5 m for the lightest (but less effective) *hasta velitaris* type and 34.8 m and 33.7 m for the barbed Talamonaccio and the pyramidal Renieblas types respectively. In contrast, the *Deva* 'tests' were a result of impromptu training, using relatively unfit, mostly middle-aged re-enactors (not, as they would argue to the contrary, prime physical specimens). Initial throws, using *pila* of pyramidal-type heads, similar to Renieblas (but with slightly shorter shafts: perhaps two-thirds length, similar to Smihel type 3 in the Connolly tests), achieved disappointing distances of between 5 and 15 m, falling short of the target. After goading them by saying that they would never hit anything (not even the proverbial 'barn door'), substituting the target with the centurion's freshly painted new shield, to his disappointment distances of around 20–25 m were achieved, the majority hitting the target, most fully penetrating and effectively destroying the shield (those *pila* remaining embedded in the shield then being difficult to remove, causing considerable damage on extraction).

Factors to consider in these two tests are:

Connolly test:
• Used accurately reconstructed *pila*, manufactured specifically for the test.
• Tested under controlled conditions using scientific measurements.

- Used an athlete who was an experienced javelin thrower.
- The test subject was probably emotionally calm, rather than adrenaline fuelled as under combat conditions.
- The test was conducted using a target of flat-profile plywood, so did not experience the potential of 'tangential' strikes.

Deva test:
- Used *pila* purchased for re-enactment use, as accurate reconstructions, but not manufactured specifically for the purpose of the experiment and lacking any known historically comparable metal qualities.
- Tested under impromptu conditions, without accurate measurement.
- Used relatively unfit, middle-aged re-enactors.
- Used test 'subjects' who were not emotionally calm, but adrenaline fuelled – not to the level of battle conditions, but through goading for a 'dare' and the pure 'mischief' factor of showing off.
- The test was conducted using a target of a reconstruction semi-circular Roman shield, which, although possibly over-engineered, and not constructed of entirely accurate materials, would have offered a more realistic target, offering the potential for 'tangential' strikes against its curved surfaces.

The *Deva VV* test findings are not, however, at variance with those of Connolly, who similarly reported satisfactory penetration with his pyramidal-pointed *pila*. It was only with the barbed heads (particularly the large ones) that he reported poor penetration results against the target, leading to his difficulty in understanding the function of these barbed-headed types (Connolly, 2000, 46), although shields were not the only possible targets that *pila* may have combated.

In the *Deva* test, no *pila* were found to have been bent on impact with either the shield or the ground. Connolly similarly reports few *pila* bending on impact with the target, although some did so on hitting the ground. He also reported no bending problems in separate 'point-blank' testing (at around 5 m distance from the target), concluding that 'although many *pila* may have been bent in the heat of battle, this could not be the prime function of the weapon' (Connolly, 2000, 46). However, the purpose of this discussion is to consider the function of the shield, rather than the weaponry, and it is clear from both tests that the *pilum* (at least those with pyramidal heads) would have seriously compromised the effectiveness of a shield, either by complete penetration and possible wounding of the user, or by partial penetration, the embedded shafts rendering the shield un-wieldable, requiring the user to discard it (not ideal under battle conditions). Furthermore, whereas moderate arrow damage to a shield could be repaired after the conflict by application of patches, the damage caused by *pila* would be more substantial, probably requiring total replacement (with possible reuse of shield furniture, such as bosses and metal rim edgings).

DISCUSSION

'Field testing' of the reconstructed shields found that even substantial three-ply shields, when provided with minimal covering, were easily penetrated by arrows at relatively close quarters (even those arrows without metal tips) and saw a high level of damage in light combat situations (even where just using wooden weapons combined with the assailant's body weight). Clearly the advantage of building shields of greater thicknesses would have been offset by the disadvantages of increased weight and lack of manoeuvrability. The addition of light padding to the shield surface permitted the use of lighter, thinner shields, built of fewer ply layers, the increased padding serving (as with the padded undergarment combined with mail/scale) to absorb the energy of the blow, slowing and reducing its penetration. It was clear that, although a shield may have survived for long periods if fortunate enough not to see combat, under combat situations it would have been seen as a very short-life commodity (with the estimated lifespan used in chapter XI possibly being overly generous in less peaceful periods). However, by use of a softer, padded outer layer, impact damage could have been minimised (also reducing splintering at penetration points), with some minor damage being repairable by patches to increase its useable life, in order to see secondary combats.

Bow strength (lbs)		154	80	70	60	40	35
Force required to draw bow (Ns)	F	700	364	318	273	182	159
Distance bow is drawn (m)	x	0.58	0.58	0.58	0.58	0.58	0.58
Estimated efficiency of bow = 0.9	e	0.9	0.9	0.9	0.9	0.9	0.9
Mass of the bow (kg)	M	1.0	1.0	1.0	1.0	1.0	1.0
Factor representing the kinetic energy of the bow (estimated)	k	0.07	0.07	0.07	0.07	0.07	0.07
Mass of arrow (kg)	m	0.06	0.06	0.06	0.06	0.06	0.06
Acceleration due to gravity (ms^2)	g	9.81	9.81	9.81	9.81	9.81	9.81
Speed of arrow (ms) (eFx / (m + kM))$^{-2}$	v	53	38	36	33	27	25
Maximum distance at 45° (m) (v^2 /g)	d	287	149	130	112	74	65

Table 7. Comparative calculations of arrow speed and maximum distances for a range of bow strengths (based on Rees, 1995).

Arrow	Padding alone		Un-padded mail		Lightly padded mail (leather) (2 mm)		Medium padded mail (4 mm)		Heavily padded mail (18 mm)		Wooden shield 3ply (10 mm)
Penetration compression	P	C	P	C	P	C	P	C	P	C	P
Short range – 9.23 m (30 feet)											
1 (untipped)	0.0	0.0	0.0	0.0	0.0	0.0	0.0	0.0	0.0	0.0	45.0
2 (needle bodkin)	9.4	0.0	10.7	0.0	10.7	0.0	7.1	3.0	3.0	0.0	full
3 (light bodkin)	0.0	3.6	4.8	0.0	4.3	0.0	0.0	4.6	0.0	0.0	full
4 (heavy bodkin)	9.4	0.0	9.7	0.0	9.7	0.0	7.1	3.0	2.5	1.0	full
5 (leaf bladed)	3.3	2.0	6.6	0.0	4.3	0.0	3.3	2.5	3.3	1.5	full
6 (trilobate)	9.7	0.0	8.1	0.0	7.1	0.0	7.6	0.0	3.6	0.0	full
Longer range – 18.28 m (60 feet)											
Penetration compression	P	C	P	C	P	C	P	C	P	C	P
1 (untipped)	0.0	0.0	0.0	0.0	0.0	0.0	0.0	0.0	0.0	0.0	15.0
2 (needle bodkin)	2.3	0.0	2.7	0.0	2.7	0.0	1.8	0.8	0.8	0.0	24.0
3 (light bodkin)	0.0	0.9	1.2	0.0	1.1	0.0	0.0	1.1	0.0	0.0	22.5
4 (heavy bodkin)	2.3	0.0	2.4	0.0	2.4	0.0	1.8	0.8	0.6	0.3	24.0
5 (leaf bladed)	0.8	0.5	1.7	0.0	1.1	0.0	0.8	0.6	0.8	0.4	21.0
6 (trilobate)	2.4	0.0	2.0	0.0	1.8	0.0	1.9	0.0	0.9	0.0	22.0

Table 8. Arrow penetration test. Shooting a selection of arrow types, at reproduction mail and shield, using varying levels of under-padding, at a range of distances (average penetration and compression in cm).

X

THE SHIELD IN USE

Previous chapters have discussed the historical origins and development of the Roman shield, how it was constructed (including performance, resistance to damage, rate of repair/ replacement, etc.), and elements of decoration, possibly for purposes of unit identification. However, these also raise questions of how the shields were actually used by their owners, and how this interrelates to construction materials and development over time.

Clearly each shield was used by an individual, but that individual operated as part of a larger body of men. Although he may have on occasions been required to fight as an individual, at other times he may have functioned as part of the larger collective body, whether that was just a small contingent (in the Roman period, perhaps just a *contubernium* of eight men) or perhaps larger groupings (where formations of a large body of well-organised men would have the additional psychological advantage over enemy troops).

While contemporary texts may provide us with some insight into the individual and group uses of the Roman shield, the most helpful source of information is probably from visual representations: those on grave *stelae*, usually of static individuals standing to attention, perhaps being the least helpful; whereas those on monumental sculptures, engravings and other decorative artistic imagery may depict the bearers in combat stances. In addition, sculptural sources can provide images of shields in use from pre-Roman periods, some possibly indicating prior stages in development, although others more likely just indicative of similar functions being applied to similar objects (not necessarily of direct ancestry to 'Roman' shields), but again demonstrating both individual and collective uses in combat situations (some of which are described below).

SHIELD WALL FORMATION

Tacitus described the use of the shield wall by legionary soldiers in close-order fighting, being used to bunch opponents into confined spaces (*Hist.* 2.22; 2.42). However, the concept of warriors forming up in a line behind a solid 'wall' of shields may have a long-standing history much before its documented use by the Roman army, although the archaic 'shield wall' formation may have been entirely different in nature to the later incarnation of the Roman period.

One early visual representation of how shields may have been used by a collective body of men can be seen on the Vulture *stela* from Telloh (*c.* 2500 BC), which shows a group of armoured soldiers belonging to Eannatum of Lagash, each carrying a spear, advancing in an early 'shield wall' formation, behind large rectangular shields (Fagan, 2004, 85; Fig. 27). However, the image is enigmatic in that it appears to show four shields, with nine heads visible above and eight pairs of feet below, and with six right hands carrying spears appearing to the side of each shield. Clearly this is not suggesting that ancestral warriors had more than the usual provision of one left and one right arm each, so must be an artistic convention representing further rows of warriors in close formation to the rear. This could also account for the multitude of heads and feet exceeding the number of shields visible. However, it is also possible that these were particularly oversized shields (depicted as reaching from shoulder to ankle), which may have been doubly wide, requiring perhaps two men to carry each. This would then suggest a very different, unwieldy method of use compared to the later Roman 'shield wall' formation.

In appendix I, while describing the early history of the Roman army, the influences from Greek and Etruscan fighting styles and armour panoplies are discussed. It is known that from the eighth century BC, a new form of armour and battle formation was introduced in Greece (the close-formation 'hoplite' phalanx of heavily armoured soldiers).

HOPLITE TRAINING AND THE USE OF SPEAR AND SHIELD

Xenophon wrote about training and discipline in the use of the spear and the shield (*Hell.* 2.4.6; *Mem.* 3.12; *Lac. Pol.* 11.5–10; Tuplin, 1986, 37–66). Based on sculptural evidence and the writing of ancient sources, such as Xenophon, Herodotus and others, there were four main basic movements used by Greek Hoplites (Connolly, 1981, 42–43):

1. When 'at ease', the hoplite stands with his spear leaning on the ground and his shield resting against his thigh. According to Connolly (*ibid.*), the hoplites sometimes retained this position in the face of the enemy as a sign of contempt. However, there is no direct evidence for this. Many artistic representations show the shield being rested on the ground and held by the left hand in readiness.
2. When called to 'attention', the hoplite may have brought his spear up to his right shoulder and lifted his shield up to cover his torso.
3. From the 'attention' position, the hoplite may have come to the 'on guard' or 'stand to' position, which meant the hoplite bringing the spear forward until his right arm was straight, with his spear being held parallel to the ground at waist level. This was the position of the 'underhand thrust'. It is believed that this was the stance in which the hoplite would advance towards the enemy. According to Connolly (*ibid.*), the 'underhand thrust' could not be executed under normal conditions in close array, as one would have to open up the 'wall of shields' in order to 'thrust' at waist level. Also, there would be a risk of striking the next man in file with the pointed butt of the spear.
4. The normal striking position has to be with the spear raised above the right shoulder, striking downwards at a slight angle, through the dart-shaped gap between the top of one's own shield and that of one's neighbour on the right. The angle of the stroke should ensure that it passes over the head of the next in file. To take up this attitude from (2) would be impossible, as it would leave the spear butt pointing forward. To do it from (3) is equally impossible with a long spear. It would be necessary to raise the spear above the right shoulder and then reverse the grip. These movements performed within the phalanx would require considerable training. A professional soldier could always tell a badly trained army by the untidy way in which they performed this drill (Connolly, *ibid.*).

From around 650 BC, similar 'hoplite' panoplies are found in Etruscan contexts, with Rome and other Latin regions following soon after, although Rich (2007, 18) suggested that they may have adopted the equipment without necessarily using the phalanx formation, continuing to fight in more flexible formations, with a mixture of heavily and lightly armed troops. The 'Phalanx', however, is still a form of 'shield wall', which will have had some formative influence on the looser 'manipular' combat formation seen from the fourth century BC.

In the 'Phalanx', hoplite soldiers fought in close formation, each man protecting the right side of his neighbour, in rows between eight and forty men deep (Goldsworthy, 2003, 22). Each hoplite was equipped with bronze helmet, cuirass and greaves, carrying a spear, a short slashing or thrusting sword and a circular shield of around 90 cm diameter known as a 'hoplon'. Popular from eighth to seventh century BC (in its earlier Greek form it was bossless), in one incarnation it is also known as the 'Argive' shield (relating to Argos, the city in the ancient region of Argopolis, Peloponnese, Greece), and is also known in Greek as the '*Apsis*'.

The shield was carried on the left arm, gripped by passing the hand and forearm through two looping straps on the reverse. The phalanx was formed of men standing side by side in close formation with shields overlapping, in a bank of men eight or more rows deep, their shields protecting the user and the man to his left. In battle they would advance as a block using their combined weight to push back the enemy (as in the later 'shield wall' formation). The weakness of the formation, however, would have been on the right side of the phalanx where the end man was not fully protected by his neighbour. Also, it is suggested that in use the body of men in the phalanx would tend to drift to the right, as there was a temptation, however unintentional, for men to shelter behind the shield to their right. This tendency was recognised and combated by the use of more experienced men to the right of the formation. The fact that this formation remained in use for so long, being adopted by numerous armies of other Latin peoples, indicates that it was clearly effective and solid against a block of enemies, but restrictive on the ability of the individual member to fight effectively. The Gallic incursions at the end of the fourth century BC introduced the Roman army to an enemy fighting in more flexible, open formations against which their rigid phalanxes struggled. It is suggested that it was at this time, in answer to this open enemy battle formation, that the Romans introduced the looser, manipular formation, whereby the body of men was broken up into 'maniples', or 'handfuls' (Keppie, 1984, 19).

The longer '*scutum*' shield, covering a greater part of the user's body, had been introduced to the Roman army possibly as early as the sixth century BC (according to Connolly, 1981, 95), this being attributed to the Servian reorganisation of the Etrusco-Roman Army, or perhaps in the late fifth and early fourth centuries BC (according to Gunby, 2000, 362), reintroduced to the local population by Celtic groups from central Europe moving south. She also supported Stary (1981) and Eichberg (1987, 178–181) in claiming that the '*scutum*' may have remained in use in the peripheral areas of Etruria and the Po Valley, where Greek influence was weaker. The wider readoption in the fourth century BC (along with the development of manipular formation) she saw as due to the 'demands made upon the Italic armies by the Celt's military strategies' (Gunby, 2000, 362). In any event, descriptions of the shape of the fifth-century BC '*scutum*' appear in both Virgil (*Aeneid*, 8.662) and Livy (*History of Rome*, 38.17), where it is named as the '*Scuta longa*', and by the fourth century BC the Roman army had equipped at least some soldiers with the longer '*scutum*'. This was not only better for individual combat situations (e.g. skirmishing), but also offered protection of a greater area of the user's body when used in 'shield wall' formation.

The fully developed shield wall formation, as used by the Roman army of the Imperial period for both offensive and defensive situations, would have presented an enemy with a solid continuous line of men in close formation, protected shoulder to knee by the curved, rectangular '*scuta*' (eliminating any gaps seen in the earlier oval Republican-period shields), the only parts of their bodies exposed being the head (protected by helmet) and below the knee (protected, in the case of heavy infantry, by leg greaves). The curvature of the shield would allow better overlap with neighbouring shields, with the added strength and deflective properties afforded against projectiles striking their surface at a tangent.

The efficiency of this formation, in its fully developed incarnation, can be seen in its continued use by modern riot police for crowd control, where, using shields similarly shaped to the Roman *scutum*, troops can advance with maximum protection while pushing opponents back with the minimum of force (and ideally without the use of weapons). I have also personally witnessed very effective 'shield wall' formation being used by Roman re-enactors during filming for a documentary about the Augustan-period '*Varusschlacht*' of AD 9, where, much to the director's dismay, despite every effort to 'lose', the 'Roman' contingent repeatedly defeated their *Cherusci* opponents, pushing them back with a solid wall of *scuta* alone (with other weapons safely sheathed).

In practical application the shield wall was not always exclusively used in a linear formation. It could also be adapted to combat situations; sometimes offensively to form a V-shaped advance to cut through enemy lines; or defensively, to form protective squares, with shield walls lining the edges.

THE '*TESTUDO*' FORMATION

A further very well-known adaptation from the 'shield wall' base, the '*Testudo*', or 'Tortoise' formation was used on numerous occasions for both offensive and defensive situations, and can be evidenced not only in literary descriptions (for example Ammianus Marcellinus described the use of the *testudo*: 6.12.14; 20.11.18; 26.6.16), but also from sculptural representations.

As shown on Trajan's Column (Fig. 69), soldiers in *segmentata* (generally identified as legionaries, but possibly representing units of heavy infantry, legionary or auxiliary) are depicted assaulting a besieged enemy while sheltering under *testudo* formation of shields. The front row of shields can be seen to overlap those behind, with a further row of vertical shields raised along the sides, enclosing the formation. The block of shields is then raised higher at the front, so that objects thrown down from above will slide over the *testudo*.

Similarly, on the Column of Marcus Aurelius (Fig. 70), soldiers (wearing *hamata*, so, depending on interpretation, possibly auxiliary or possibly just less heavily armoured

Fig 69. *Testudo*: as portrayed on Trajan's Column. ('Storming of a Dacian Fortress', from Cichorius, 1896, *Die Reliefs der Traianssäule*, plate LI)

legionary infantry) can be seen forming a *testudo* to protect them from assault by an enemy on the wall above, who can be seen throwing down objects (rocks, spears, pots of liquid, burning torches, and round objects resembling either wheels or shields). As with the previous example, the *testudo* can be seen to have been formed with the front row of shields overlapping those behind, in order that the projectiles cast down could be directed safely over the formation.

In the study of the *testudo* formation, the participation of living history re-enactment is invaluable, testing the sculptural imagery with 'real' subjects and equipment. For example, some practicalities of use, 'strengths' and 'weaknesses', can be brought to light, as seen in the modern-day reconstruction of the *testudo* as portrayed by the Ermine Street Guard living history group in recent years at Caerleon, although on that occasion with the formation demonstrated on flat terrain rather than as a sloping assault. In their portrayal they used the front row of men to form a 'shield wall', with sides similarly protected by vertical shields held by men at the end of each row behind, and with the 'roof' of the *testudo* then formed by the body of men inside the block.

Fig 70. *Testudo*: as portrayed on the Column of Marcus Aurelius. (Artwork by J. R. Travis)

Fig 71. *Testudo*: as displayed by members of Legion Ireland. (Photograph by Marcus Vlpivs Nerva (Martin McAree) of Legion Ireland)

However, in contrast to the two sculptural examples, and the below portrayal by the Legion Ireland Roman group (Fig. 71), the 'roof' has been formed by overlapping the rear shields over those in front. As can be seen on the photograph, this creates a gap through which projectiles could enter the formation from above, with further gaps between the front row and roof exposing the faces of front row troops (which would have been reduced by using the previously described frontal roof overlap, and consequential slight downward 'tipping' of the front roof edge). Similarly, gaps are formed between the roof and sides, as these shields stand proud from the protective cover. However, the latter 'weakness' would probably be of lesser concern than the former two problems (the block being less likely to receive lateral than frontal and aerial assault), and may have been an acceptable risk. It would never have been possible to make the block fully impenetrable, the aim being to minimise losses. It would seem plausible to suggest therefore, that the formation should always strive for overlapping of front rows with those behind, as depicted in the sculptures, rather than 'underlapping' as seen in the modern Ermine Street Guard reconstruction.

USE OF THE SHIELD IN 'INDIVIDUAL' COMBAT SITUATIONS

Although we tend to think of the Roman army as fighting as a solid, collective body of men, in battlefield contexts of one major army against another, each individual soldier would also have had to be prepared to fight as an individual in 'one-on-one' situations, in much the same way as any preceding or subsequent combatant. However, in the same way that different fighting styles evolved during the history of the Roman army, different fighting styles were used by earlier warriors and opponents encountered. The shield, while having been used from very early periods, may not always have been purely combative, serving a ceremonial purpose in some cultures, or may have only been used by a limited number of combatants, possibly based on status. Some shields, therefore, may not have been of a construction substantial enough to withstand heavy assault, or may have been excessively heavy, impeding effective use in skirmish situations. However, in general it should be assumed that the primary function of a shield has always been combative, to deflect an attack if necessary (although any warrior in a combat situation will make creative use of any piece of his equipment, for both defensive and offensive purposes). This is highlighted by the many highly decorated, high-status Germanic/Roman shields found among the finds from Illerup, some with evidence of repaired combat damage (Ilkjaer, 2002, 99; Jørgensen, *et al.*, 2003, 313; Fig. 42), and the legionary shield from Dura Europos (James, 2004, 182), all of which could have equally served ceremonial or combative functions.

The long, oval 'La Tène' or 'Celtic' shield, suggested as being ancestral to the Roman '*scutum*', may not have been carried by all Celtic combatants, but was clearly seen to be effective in use for individual skirmish fighting to such extent that it was adopted into the Roman panoply. Being constructed of planks, rather than ply, it would have been heavier to wield but more substantial against an assault, inflicting more damage on an opponent when used offensively. The ply shields, however, when strengthened on the edges with a metal strip (a practice which also appears to originate in northern regions), would also make for an effective weapon when used to strike upwards into an opponent's face, or downwards into the foot or shin. Similarly, the metal shield boss (or the earlier wooden boss, strengthened by a sheet metal covering) could also be used as an effective weapon, being used to strike in close-quarters fighting, as described by both Tacitus (*Hist.* 4.29; *Agric.* 36) and Livy (9.41.118).

'ROMAN' USE OF SHIELD: FIGHTING STYLES, FUNCTION OF INDIVIDUAL

Trajan's Column (Fig. 7) shows an advancing body of men in *segmentata* (nominally attributed as 'legionaries', but not necessarily so) and *hamata* ('auxiliaries'), carrying rectangular and oval shields respectively, shields in their left hands, swords in the right. On one figure the rear of an oval shield can be seen, with the user grasping a horizontal bar, approximately a quarter to a third of the way up the shield, with the forearm passing through a strap placed horizontally across the centre (contrary to the usual view of vertical handgrips being used on oval shields). All but one of the participants are shown to be advancing, with their weight mainly on the front, leading leg. One is shown to be leaning backwards, with his weight on the rear leg, arm raised, possibly ready to throw some projectile (perhaps spear or *pilum*), which is now missing.

Further imagery of soldiers (with varying interpretation as auxiliaries and legionaries) can be seen on the Mainz pedestals (Fig. 12). These include two individuals (on Figs 12a & 12c) advancing, with their weight on their forward, leading leg, one carrying his sword in the right hand and rectangular shield in the left (Fig. 12c), the other carrying spear/*pilum* and oval shield. The oval shield appears to be flat (although the flatness may just be a convention to suit the low relief of the sculpture) and combined with a round boss. This soldier also appears to be carrying at least two more spears/*pila* behind the shield, although these do not project far beyond the shield and may be quite short in length. It should also be noted that it would have been difficult for the individual to carry both shield and spears in his left hand, so these additional weapons may have been held to the rear of the shield by some form of clips (which would have required some form of easy-release mechanism).

Further 'legionaries' can be seen depicted on the *Tropaeum Traiani* at Adamklissi with their shields being used offensively. They are shown as heavily armoured, with reinforced helmets, laminated sword-arm protectors (*manica*) and leg greaves, some in mail (Figs 8 & 9) and others in scale (Fig. 10), all carrying curved, rectangular shields in their left hand, sword in the right hand, fighting Dacian opponents armed with long, curved *falx* weapons. The individual in Fig. 8 can be seen using his shield to push the enemy over, while striking down his opponent with a raised sword. The individual in Fig. 9, however, uses his shield raised to face height, in order to strike his enemy in the face with the shield, while simultaneously delivering the fatal blow upward into the abdomen of his opponent, bringing his sword up under the lower edge of the shield. As they are all acting 'offensively', all three legionaries can be seen to be using an advancing stance, with their weight mainly on the forward, leading leg.

A further three legionary figures can be seen on the *Tropaeum Traiani* (Fig. 11) who, although not appearing to be advancing or actively engaging in combat, do provide valuable additional information on the use of their shields. The figures are stood in line, side by side, with drawn swords in their right hands. They wear long-bodied *hamata* (which appear to have sleeves of at least short length), reinforced helmets and leg greaves. They carry curved, rectangular shields in their left hands, which appear to be pushed to one side by the left shoulder, exposing an additional carrying aid – the shields being suspended by wide, buckled shoulder straps crossing the bodies from the right shoulder to under the left arm. Their left hands are not visible but can be presumed to be grasping a rear handle, although it is not possible to determine if this would have been of vertical or horizontal orientation. However, on curved, rectangular shields, these are generally accepted as being horizontal, as on the example from Dura Europos (Fig. 40).

The figure of a cavalryman seen on the Arch of Galerius (Fig. 73), wearing a belted, loose-fitting scale coat and *spangenhelm*-type helmet, can be seen carrying a round (or wide oval) shield, possibly similar in shape to those from Dura Europos. Here, unlike many sculptures, the rear of the shield is visible, showing the method of carrying. The user can be seen to have his forearm passed through a centrally placed vertical strap, grasping the shield using an asymmetrically placed vertical handgrip, reaching from one edge to the other, placed one-third to one-quarter of the way across the shield (closest to the right-hand side, as viewed from the rear). This asymmetrical suspension would have caused the shield to hang in such a way as to naturally cover more of the side and rear than the front of the user's body (where assaults would most logically originate during cavalry use), while not encumbering the user's position when riding. The double-arm-strap-and-handle method of grip allows greater manoeuvrability in use, with the user also able to release the front handle without losing his shield, in order to control his horse yet still have his sword arm free for use.

In contrast, a panel from the Arch of Marcus, reused on the Arch of Constantine (Russell-Robinson, 1975, 86, plate 241), depicts legionaries, Praetorians and auxiliaries from the late second century AD carrying shields which appear to be oval, but lightly curved (not flat). The rear of the shield is visible (so by consequence, identification of shape of boss or any frontal decoration is not possible), but from the user's stance appears to be held by a horizontal, not vertical, handgrip. This suggests that it would be wrong to blindly assume that all oval shields used a vertical grip.

On a relief from Osuna in Spain, dating to the first century BC, two legionary figures can be seen (Fig. 72), one at least wearing *hamata* (the armour of the other figure being unclear), both wearing greaves. Both are in advancing stance, their weight on the forward, leading leg, although stood back to back, pointing in opposite directions, one with his sword drawn in his right hand. Both figures carry long shields, with oval *umbo* and a long *spina* (as on the Fayum shield) and a wide edging strip. Their shape appears to be slightly curved, straight sided with rounded corners and slightly rounded top and bottom edges. The shields are neither the conventional oval nor rectangular shape, but appear to be wider at the top than the bottom, with sides tapering gently towards the base (not unlike the medieval 'pavise'). This unconventional shape may possibly be an indication of a variety of shapes being in use, perhaps denoting different units or different functions within a unit.

It would seem reasonable, therefore, to also propose that there not only appear to have been a variety of shield shapes in use simultaneously, serving a range of purposes, some being combative (such as heavy infantry, light infantry, skirmish troops, cavalry, etc.), others being more ceremonial (for example as used by standard bearers), but that these appear equally to have used a variety of different handgrips and carrying methods, each as was best suited to its intended purpose and fighting style.

Fig 72. Relief of legionaries from Osuna, Spain, first century BC. (Artwork by J. R. Travis, from Russell-Robinson, 1975, 164, fig 175)

Fig 73. Cavalryman on the Arch of Galerius in belted, loose-fitting scale coat, wearing *spangenhelm*-type helmet. (Artwork by J. R. Travis, from Russell-Robinson, 1975, 93, fig. 123)

XI

REASSESSING THE EVIDENCE

From the sculptural representation, literary descriptions and the few physical remains of shields discovered through archaeological methods, a developmental progression can be seen through the history of the Roman army, from its conception through to the end of the Empire. During this time it underwent a series of changes (at times radical), from the round '*hoplon*' into the long 'Republican' shields with curved sides and ends (like the Fayum example), then into the semi-cylindrical, rectangular 'legionary' *scutum*, as seen on Trajan's Column. These shields appear, based on the Fayum and Dura complete shield examples (Kimmig, 1940; James, 2004, xxix, plate 10) and based on the description of Polybius (6.23.3), to have been made from wooden strips, laid at 90° to each other to form plywood. The burnt remains of the Doncaster shield also appeared to have been of similar plywood construction (Buckland, 1978, 251).

However, our knowledge of constructional methods was scant until the recent publication of the finds from Masada (Stiebel & Magness, 2007, 16–22). The Fayum example was published many years ago with little information as to its construction (Kimmig, 1940; Fig. 32). The Dura example had been overzealously conserved since its discovery and any exploratory examination of the structure would cause further damage (James, 2004, xxix, plate 10; Fig. 40). Worse still, the Doncaster example had been almost completely consumed by fire, the residual charred remains suggesting in places the possibility of layered ply construction, but with its true shape and dimensions indeterminate (Buckland, 1978, 251). The resulting reconstruction by Buckland was therefore based on comparatives with other existing shield remains, Roman and otherwise, drawing analogies to the oval-shaped, plank-built Celtic shields and Iron Age shields (as seen at Hjortspring; Crumlin-Pedersen & Trakadas, 2003), leading him to suggest the possibility that the shield may have belonged to an auxiliary cavalryman, following his native traditions, or may have been a trophy of war from a previous military action.

Buckland's reconstruction also followed what has almost become an accepted convention, that the shield board should be formed from three layers of ply strips (although his choice for the board to be 'flat' rather than 'curved' seems to have been based on simplicity of production of the reconstruction). However, early interpretations of the Doncaster remains were reportedly of curved profile, similar to the Fayum and Dura shields. Furthermore, the description of estimated board thickness (and direction of residual traces of ply) on the reverse of the boss and handgrip assembly, suggests a possibility of the shield being formed from only two layers of ply, with a horizontal handgrip (as on the Fayum and Dura examples). This then conforms more to the accepted standard for what is perceived as a 'legionary' rectangular shield, rather than an oval 'auxiliary' shield.

The recently published military equipment from Masada, however, suggests that the majority of shields present may have been of only two-layered ply construction, being considerably thinner than had previously been thought. The perception held until recently of Roman shields (and as portrayed by Roman re-enactment groups) is for shields formed from at least three layers of ply (with four layers used by one group), producing an extremely heavy shield with a thickness of at least 10 mm (despite Polybius' description of only two layers used; 6.23.3).

The implication of the Masada remains, however, suggests that this view should be reconsidered, with shields of perhaps only half of this thickness, at least by the time of the

Masada siege during the early Imperial period. This may therefore be an additional developmental progression from the long Republican shield (like the Fayum example). The loss of the curved ends and sides may have been accompanied by an overall reduction in thickness, producing a much lighter and more manoeuvrable shield, permitting a faster response in close-quarters combat (necessary against the different types of adversaries encountered in the Gallic wars and Germanic campaigns). The lightness of these new shields would then have further implications for the response time for deployment of troops, the distances they would be able to travel and the speed at which this would be accomplished, particularly as they were now expected to carry their own equipment (as 'Marius' mules').

The decreased thickness of the Masada shields does not necessarily imply a weaker structure. Indeed, it suggests a more highly developed understanding of the technology of impact absorption than had been previously perceived. Where in modern-day body protection (such as bullet-proof vests) we utilise layers of soft materials, and in the design structure of motor vehicles and motorway barriers we incorporate 'crumple zones', with the intention of absorbing the impact rather than withstanding it, we find that this concept was already well understood over two millennia earlier. The fragments of the thin ply shields from Masada were found to have been augmented by the addition of flexible layers of woven textiles, covered by thin leather or parchment. This not only consolidated the ply structure, but also provided a strong, flexible, weatherproof surface which would absorb the impact from a sword or projectile strike. This would be more likely to have bounced off the surface than to have penetrated to the underlying wooden structure, or at least may have prevented full penetration in 'arrow storm' situations.

Another implication from the Masada remains lies with the later transition from the rectangular *scutum* to the wide, oval (almost round), plank-built shields of the later Empire. These are known to have been commonplace in the later Empire, with finds of fifth-century AD late Roman-style deposits at Illerup Ådal, Denmark (Ilkjaer, 2002). There are also several well-preserved examples found at Dura Europos, suggesting that this type was already well established by the mid-third century AD (James, 2004). However, if the fragments of planks from Masada are correctly interpreted by Stiebel as being from this type of shield (Stiebel & Magness, 2007, 79, plate 15), this would push back its use by a further two centuries, indicating its simultaneous use alongside the generally accepted 'legionary' *scutum* for a considerable length of time.

Its use cannot, therefore, have been entirely auxiliary, but must have represented a type of equipment issued to troops serving a different type of function within the legion, perhaps the former being preferable for heavy infantry use and shield wall/*testudo* formations, with the latter being better suited to close-quarters skirmishing and/or horseback use.

CONCLUSION

The three examples of semi-cylindrical plywood shields (*scuta*) cited, from Fayum, Doncaster and Dura Europos (Kimmig, 1940; Buckland, 1978; James, 2004), could be seen to reflect the developmental stages as seen in Fig. 1, the changing shape of the scuta over time being, in all probability, adaptation to changing enemy weaponry and fighting techniques, in much the same way that the *scutum* had been adopted centuries earlier. Similar developmental progression can also be seen in shields for cavalry use, although cavalry equipment in general (consisting mostly of non-citizen, auxiliary units), varies considerably due to the wide range of ethnic influences involved (Fig. 74).

As described, the use of long, rectangular shields for full-body protection in shield wall formations is known from at least 2,500 BC, as depicted in the Vulture *stela* from Telloh, although this does not appear to be a direct ancestor of the Roman *scutum* (Fig. 27; Fagan, 2004, 185). The *scutum* appears rather to have been adopted from the long, oval form in use by the Samnites in the fourth century BC, which was in part influential in the change to manipular battle formation. The source from which the Samnites derived their design is

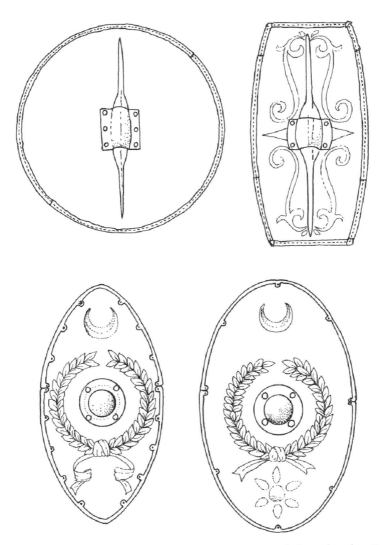

Fig 74. Progression of cavalry shield shapes, from Republican period through to late Empire. (Artwork by J. R. Travis)

unclear. It does not appear, however, as though they can be credited with its development with certainty, as markedly similar shield shapes can be seen in use in earlier Iron Age contexts in Celtic regions of northern and western Europe with similar long, central *spina* and wooden *umbo* (Buckland, 1978, 262–3).

In his discussion on the development and distribution of the typical 'Celtic' shield, Stary (1981, 297), saw the origin of oval shields with spindle-shaped boss in the early Iron Age Villanova culture of the Apennine peninsula. These then went out of fashion in Etruria at the end of the eighth century BC due to military influences from the Eastern Mediterranean, but in peripheral regions their use continued, particularly north of the Apennines. The Celts then adopted their use through cultural contacts in northern Italy and the Alps. They then brought them back with them during the invasions of Italy (fourth to third century BC), introducing them to their Samnite allies and seeing them adopted in response by the Romans (along with the adoption of manipular formation) as being superior against more 'unpredictable' enemy tactics. Etruria also adopted some of the Celtic assemblage (particularly the increased-visibility helmet) and partial use of the oval shield for some lightly armed warriors, while still retaining their hoplite phalanx formation (Stary, 1981, 297).

The developmental changes then seen on the shield design over time are twofold. Firstly, the central *spina* and *umbo* shape evolve, from a solid wooden boss covered by a simply moulded bronze sheet (as on the Fayum example) to the round, fully shaped bosses of the Doncaster and later shields, which appears to follow a Gallic influence, using methods of manufacture that has much in common with their similarly well-developed skills in helmet production. Secondly, the shield-board shape evolved, initially truncating the curved top and bottom edges, and later also squaring off the sides. This permitted a more efficient interlocking of boards during shield wall and *testudo* formations, eliminating the danger of gaps between shields (Figs 69–71). Again, these adaptations appear to coincide with changing enemy methodology, for example with campaigns into hilly northern regions against Germanic opponents and the prospect of assault by overhead bombardment with projectiles and heavy objects.

The later abandonment in the third and fourth centuries AD (although perhaps only partial) of the rectangular *scutum*, in favour of the large, plank-built, oval shields, as seen at Dura Europos (Figs 37–38; Fig. 76; James, 2004), and on the Arches of Diocletian (Fig. 76; Stephenson, 2001, 35, Fig. 9) and Galerius (Fig. 73), could again be a response to changing enemy tactics, where fast-moving manoeuvrability took precedence over the more solid, static shield wall formations. These later shields also represent a change in construction methods, moving away from the use of layered ply in favour of thicker planks. This in one respect echoes the ancestral early Iron Age plank-built shields of northern and western Europe, although these used considerably wider and more crudely formed planks. This may be indicative of a changing focus of influence, adopting methods of construction from different regions (not necessarily northern Europe), or could simply be the most efficient method to produce a convex form, curved in two planes rather than one as on the earlier *scutum*.

Fig 75. Part schematic of the 'Battle of Ebenezer' fresco from the Synagogue, Dura Europos, depicting infantry soldiers in loose-fitting, long-sleeved, knee-length mail or scale coats, some bare headed and some wearing coifs, carrying wide oval shields. (Artwork by J. R. Travis, from James, 2004, xxvi, plate 4)

The examples cited do, however, show a continuity of methodology, in the application of surface coating on the long, oval, semi-cylindrical and later oval forms in the use of either fabric or skin, sealed by a painted outer surface. This served both protective and identification purposes, while helping to consolidate the physical structure. This flexible painted outer layer would not only have protected the wooden surface from the elements (although removable leather covers are also known to have been used) but would also have increased the impact absorption properties of the shield, increasing its useful life. While these surface coatings, leather or fabric, would probably have been normally applied as a single sheet, the use of recycled materials cannot be ruled out, with the possibility of the use

Fig 76. Late third- to early fourth-century AD armoured infantrymen from the Arch of Diocletian, Rome, wearing long-bodied, long-sleeved, loose-fitting mail and scale coats. (Artwork by J. R. Travis, from Stephenson, 2001, 35, fig 9)

of patches of leather from old tent panels, or strips of fabric from worn-out tunics. Strips of fabric were known to have been used in the construction of Renaissance-period panel painting (a process with many features in common with Roman shield production). However, although vegetable-based fabrics, such as linen, would be useable in this way, wool-based fabrics would not, due to their tendency to shrink when wet, which could cause separation and 'gapping' between strips. As military tunics were normally made of wool, these would not be useable (unless shredded and processed into 'felt'), but linen undergarments or civilian clothing could possibly have been recycled in this way.

Shields were nevertheless a short-life piece of equipment (in comparison to helmets and body armour, which may have passed down to multiple owners) and would have required continuous replacement. Although Pliny (*Hist. Nat.* 16.77) described the best woods to use for shield construction, the variety seen in the examples cited suggests compromise to the best currently available. In view of the large quantities of wood which would have been involved, it is probable that some compromise would have been necessary, to use locally available timber wherever possible, with importation from other areas when local supply was exceeded.

The life expectancy of a wooden shield (despite surface coating of fabric, leather and paint), even in peaceful time periods (with minimal combat damage), would probably only be around two years. For an average legion of approximately 5,000 combatants this would require the production of around 2,500 shields per annum, or approximately forty-six shields each week. Even if shields were to last a more generous four years, this would involve production of at least twenty-four shields per week. This would imply a high level of workshop activity in each fort being devoted to shield production alone, with processes involved consuming large areas of workshop space and storage of materials taking up more, not to mention the need for areas for construction of ply bases, presses, drying, surfacing and painting areas, none of these activities leaving any significant archaeological traces.

Furthermore, providing sufficient timber for shields was produced from each tree felled, this would have resulted in the loss of 200 to 400 trees per year for each legion (and an even greater number in less peaceful times, where production would need to increase to meet the demands of replacement from combat damage), in addition to those felled for constructional or fuel purposes. Although legions were rarely at full strength (so the estimated 5,000 combatants may be slightly overestimated), and although the narrow strips of wood used in shield production could have utilised the smaller, less useful branches of trees, or possibly even some recycled timber (perhaps explaining the use of ply rather than plank construction), these estimates are still, at the least, conservative. In less peaceful periods, shields would have required frequent repair and replacement. It is therefore possible that, to meet the high demand for timber, some level of woodland management may have been practised in the areas under the influence of each military installation (its *territorium*). Furthermore, greater use may have been made of alternative fossil fuels in order to reduce the need for wood and charcoal in manufacturing and industrial processes.

Although this level of production from established military units appears high (along with corresponding loss of adult trees), an even greater capacity would have been required for times of total re-equipping of new legions. This would, in consequence have required the felling of large areas of woodland and investment in large-scale production methods, dedicated factory sites and associated manpower, presumably, for reasons of efficiency and economy, to be situated as closely as possible to areas rich in timber. It is possible that the valuable resource of timber offered by regions such as Gaul, Germania and Britannia (in addition to those more obvious mineral resources of copper, lead, iron and coal) may also have been influential in expansionist policies of the Imperial period, in a cyclical process of expansion, requiring increased military strength, requiring additional military resources, in turn provided by expansion.

APPENDIX I

THE ORIGINS & DEVELOPMENT OF THE ROMAN ARMY

THE 'ANCIENT' PERIOD AND EMERGENCE OF THE REPUBLIC

The early period of Rome's history, from its almost mythical foundation by Romulus, supposedly on 21 April in 753 BC (or 748 BC according to Fabius Pictus), is not well documented, the known written history only commencing towards the end of the third century BC.

Initially Rome was only one of a number of small, frequently warring communities of indigenous Latin-speaking peoples of the Italic peninsula, sharing a common language, similar material culture and religion, and some common sanctuaries (Burns, 2003, 62–63; Rich, 2007, 8; Le Glay, *et al.*, 2005, 1–7). Of these, their immediate neighbours were the *Sabines*, to the north-east; the *Aequi*, to the east; *Latium*, to the south; the *Volsci*, further south (Rich reported some controversy surrounding the first appearance of the *Volsci* in the Italic peninsula; traditionally they were already in the area during the 'Royal' period, but modern scholars propose an 'invasion' in the early fifth century BC, although there is no reference to support this in the classical sources; 2007, 10); and the *Hernici*, further south-east. Yet further south and into Sicily were Hellenised former Greek colonies, which Polybius called '*Magna Graecia*', and Cicero described as 'the old Italian Greece, that used to be called Great', and these territories funnelled Greek and Etruscan artistic and cultural influence into the Italic mainland through trading activities (Le Glay, *et al.*, 2005, 14).

Their most immediate neighbour to the north (with Gallic tribes still further north), the Etruscans, were more technologically advanced, possessing systems of drainage and irrigation to improve crop yields, practices of mining tin, copper and iron, and craftsmen skilled in the production of tools and armaments, which display strong Greek stylistic influences. They also had a more highly developed political structure, which better fits the finer definitions of a 'State'. Theirs was a patrician society, with rural plebeians and a huge servile class, and it was split into a federation of twelve states, each under the control of elected magistrates but with the option to elect a single leader should events dictate the need. In contrast to their neighbours (who at that time were still living a simple, pastoral existence in village communities of small huts), they lived in cities with substantial stone buildings and temples and public spaces. They used a different language but shared some similarities of culture and religion, with a pantheon of gods similar to that of the Greeks and prototypical of the later Roman pantheon (Le Glay, *et al.*, 2005, 9).

The earliest habitation at Rome dates from around 1000 BC, developing gradually by the eighth century BC into a series of small clusters of huts on each of the seven hills (Le Glay, *et al.*, 2005, 22). In March and October of each year, rituals marked the opening and closing of a campaign season of inter-community warfare throughout the Italic peninsula, Rich (2007, 10) suggesting these conflicts were carried out for the purpose of territorial expansion (although disputing frequency of conflict), while Burns (2003, 62–63) described them as raids and skirmishes carried out by aristocratic warrior bands, fighting for personal glory and booty in the best 'Homeric' tradition.

The Etruscans similarly appear to have embarked on a policy of expansionism, leading by the sixth century BC to Etruscan rule in Rome, possibly as a result of trade, intermarriage of dynasties, or by force. Traditionally, the last three 'kings' of Rome were formed from this new 'Etruscan' dynasty (Tarquin the Elder, Servius Tullius and Tarquin the Proud, or Tarquinius Superbus), although their existence is disputed, their names possibly being no more than symbolic of Etruscan presence, influence and domination. Le Glay, *et al.* (2005, 24), however, suggested that this point in time should be better considered as the 'true foundation' of Rome, as the Etruscan influence stimulated a series of innovations (including improved water supply and drainage, substantial new public buildings and defensive walls).

Weapons and armour are found as grave goods throughout Italy from around 1000 BC (bronze armour and iron weapons), but as their use does not seem to be widespread, appearing to be mostly a high-status affair reserved display purposes, we cannot be sure how widely weapons and armour were used, nor can we adduce any information about fighting styles or battle formations (Rich, 2007, 17). It is known that, from the eighth century BC, a new form of armour and battle formation was introduced in Greece (the 'hoplite' phalanx of heavily armoured soldiers), and from around 650 BC similar panoplies are found in Etruscan contexts, with Rome and other Latin regions following soon after. The 'hoplite' soldier was equipped with bronze helmet, cuirass and greaves covering his lower legs. He carried a spear (around 2.45 m long), a circular shield (around 90 cm diameter) known as a '*hoplon*', and a short slashing or thrusting sword. They fought in close formation, each man protecting the right side of his neighbour, in rows between eight and forty men deep (Goldsworthy, 2003, 22).

However, Rich (2007, 18) cited Van Wees (2004) when he disputed that this new panoply would necessarily have led to a new style of fighting. Although in Greek use the phalanx line-up would have included men of all classes fighting together on an equal basis, aristocratic and middle class, he believes that the same may not have been the case in Etruria, where society was dominated by an aristocratic elite (*gentes*). Citing hoplite equipment found alongside traditional Etruscan weapons, he suggests that the new equipment could have been adopted without necessarily also adopting the close-formation phalanx, continuing instead to fight in more flexible formations, but with a mixture of heavily and lightly armed troops.

THE EARLY REPUBLIC

The army in the earliest stages of the Republic was quite different to that at the time of its transition into Empire. As with any human-based system, it evolved over time, reflecting the evolution of the society that it protected. It was a citizen militia, recruited and organised along parallel lines to the political infrastructure, through the *Comitia centuriata* voting assembly, categorising the available manpower into classes according to wealth and ability to provide equipment (see Table 9).

The complexity of the initial incarnation of this 'Servian' system has been disputed. Livy (1.43) described a system of five classes of varying levels of equipment, which may represent its much later structure, whereas modern historians, including Keppie (1984, 17), suggest that its earliest form may have comprised one single *classis*. By the start of the ten-year war with *Veii* (406 BC), it is believed that the second and third classes of less wealthy men were added, when the army increased from 4,000 to 6,000 men, not all of whom could afford the full hoplite panoply. Keppie (1984, 18) suggests that the long Italic shield may have been introduced at this point to compensate for the reduced level of armour. Also around this time, the cavalry may have enlarged from six to eighteen centuries, with the state providing and maintaining horses and equipment for the less wealthy additional recruits (Keppie, 1984, 18).

The Gallic incursions at the end of the fourth century BC introduced the Roman army to an enemy fighting in more flexible, open formations, against which their rigid phalanxes struggled. It is suggested that it was at this time, in answer to this open enemy battle formation, that the Romans introduced the looser *manipular* formation, whereby the body of men was broken up into *maniples* (or 'handfuls'), a change which has been attributed to

Camillus, the commander noted for his success both in the *Veii* campaigns and in countering the Gallic threat (Keppie, 1984, 19).

During the course of the next fifty years, possibly in answer to some less-than-successful campaigns, the smaller tactical units (*maniples*) gradually became more widely used. At the same time, some changes to the internal command structure were also evident. By 320 BC the army was split into two legions, with six military tribunes assigned three to each legion. By 311 BC, however, Livy (9.30.3), reports that military tribunes had increased to sixteen, with four to each of four legions (the army having therefore doubled in size), with two of the legions commanded by each of the two consuls (Forsythe, 2007, 24–42). Despite continued expansionist campaigns over the next 150 years (the *ager Romanus* expanding from 1902 sq. km in 338 BC to 26,805 sq. km by 264 BC), Polybius (6.19–29) reports that the size of the army remained steady at four legions.

In the new *manipular* formation, centuries of infantry, around sixty men in each, were organised into thirty blocks (or *maniples*) per legion, in battle forming three rows with gaps between (the *triplex acies*). At the rear, ten *maniples* containing one century each were formed of the *triarii*, the oldest and most experienced men. In front of these, in the middle row, were two rows of ten maniples, two centuries in each, comprising the *principes* (men in their prime, aged twenty to thirty years). At the front were the *hastati*, drawn from the youngest men. All would wear the standard legionary kit of helmet, bronze cuirass and greaves. When deployed, the legion would be flanked by *alae*, or wings, of allied soldiers recruited from among the other Latin peoples, subdivided into cohorts of 400 to 600 men, along with two units of cavalry, one at each end. In front, in later periods, a row of *velites* provided a lightly armed protective screen, formed from those who could only afford the very least equipment (of helmet, shield and a bundle of light javelins) and those men too young to fight in the main line (Goldsworthy, 2003, 27; Campbell, 2000, 4; Polybius, 6.19–26; Russell-Robinson, 1975, 137). Keppie (1984, 33) linked the creation of these '*velites*' (or light 'skirmishers') in around 211 BC (according to reports by Livy) to a reduction in minimum census requirement in 214 BC, which brought into the levy men who could only afford the bare minimum of equipment.

THE LATER REPUBLIC

With the prolonged campaigns of the Punic Wars and soldiers remaining in the army longer, the quality of manpower changed during the period. The longer-serving army became more efficient, better trained and possessed improved tactics, but kept men away from their other livelihoods and family estates, leading to dissent over the levy. Recruitment became difficult, as men who had already served were reluctant to participate in a second campaign.

After the Battle of Cannae against Hannibal in 216 BC a new army was needed at short notice, but there was neither manpower nor finances available to provide this. Polybius (6.23.15; 39.15) related how many of the men in the five classes (*assidui*) could no longer equip themselves. Years of continuous campaigns had drained resources, with some of the lower classes (impoverished and dispossessed of their land to the benefit of the more wealthy) now unable to meet the property requirements, decreasing the numbers eligible for military service.

To meet this shortfall, changes were instigated that lowered the age of recruits and offered citizenship to slaves willing to volunteer (Cottrell, 1992, 153). New fiscal policies were introduced to equip this new army unable to supply its own kit, along with permitting reuse of weapons and armour taken as 'spoils of war' and stored on temple walls (Cary & Scullard, 1979, 130).

A progression of political and social reforms were introduced in the following years, to address problems of discipline and training within the legions and to increase numbers of citizens eligible for service, some with varying levels of success. It was therefore an understandable precedent to the later reforms credited to Gaius Marius in 107 BC, made in answer to the demands of the Jugurthine war, whereby Gaius called for additional volunteers from the *capite censi* (who would be equipped at the state's expense), in addition to many time-served veterans (Scullard, 1970, 52–59).

While the Jugurthine problem was being resolved in North Africa, a new problem arose on the northern frontier, with threatened incursions by Celtic tribes, the *Cimbri* and the *Teutones*. Once these threats from Gallic/Celtic incursions were resolved, the whole of the Italian peninsula came under Roman control. The divisions between Latin and Roman citizen then became less defined, with many Italian communities wanting equality and full citizenship rights, leading to the subsequent Social War of 91 to 87 BC, and the Senate eventually conceding citizenship to all states south of the River Po, which had remained loyal (Campbell, 2000, 7; Keppie, 1984, 62).

In addition to modifications to equipment, improved training, discipline and morale (introducing gladiator-style training, and use of the silver eagle standard, or *aquila*, as a rallying symbol; Pliny, *Nat.* 10.16), and requiring the soldiers to carry most of their own equipment (leading to the description of '*muli Mariani*', or 'Marius' Mules'), Marius was credited with initiating a change to cohortal formation, although similar formations had been tested by his predecessors. Polybius (11.23.1; 11.33.1) related that the new cohort system had been first used by Scipio in Spain during the Hannibalic War (at the Battle of Ilpia, in 206 BC), and Livy reports its later use during the war in Spain, in 210–195 BC. Aemilius Paulus is also credited with having used a more flexible variant of the standard *maniples* at the Battle of Pydna in 168 BC, breaking the rigid three battle lines into little squares (Cary & Scullard, 1979, 159).

The pre-Marian army had been formed into legions of around 4,000 men, organised into centuries of sixty men, forming three battle lines (the *triplex acies*), split into *maniples* (blocks of 120 men, or two centuries; Russell-Robinson, 1975, 137; Caesar, *Bell. Gall.*, 1.24.2). It has been proposed that Marius restructured this system, replacing the *maniples* with the more flexible cohorts, ten per legion, each divided into six centuries of eighty men (a cohort being made up of three *maniples* of two centuries each), with each century commanded by a centurion, or *pilus prior* (the first cohort being larger than the others). Overall, with the inclusion of legionary cavalry, the legion would then have consisted of over 5,000 men (Campbell, 2000, 6).

As a consequence of the Social War (91–87 BC), whereby many Italian communities achieved equality and full citizenship rights, the *alae sociorum* ceased to exist. Recruits were now drawn from all Italy (all now being citizens and therefore entitled to legionary status). Many of those left homeless by the wars were recruited into the army (which from this point never numbered less than fourteen legions), with their equipment costs by necessity being borne by the treasury (Keppie, 1984, 69). In their place, auxiliary 'specialist' forces came to be provided by client kings. For example, Balearic slingers, Cretan archers and, by the first century, Numidian, Spanish and Gallic units replaced the Roman and Italian cavalry. However, organised, permanent, reliable auxiliaries were not found until the Imperial period (Cagniart, 2007, 88).

Whereas in the earliest periods the Roman army was raised afresh for each annual campaign season, by the later Republican period it served an entirely different function. Keppie (1984, 77) suggested that two separate functions can be identified, describing two distinct types of army: 'standing' and 'emergency'. The 'standing' army served for long periods as garrisons in the provinces, whereas the 'emergency' army was levied for a specific campaign (as with the earliest armies), at the end of which its men would be released with accumulated 'booty' and possibly additional rewards. The number of legions (and of course the number of men serving within them) fluctuated over time. By the time of the First Punic War there were still only four to five legions, but these were beginning to be retained in service (and so retaining their identification numbering), not being disbanded at the end of the season. By the Second Punic War, fourteen more legions were added: seven in 217 BC and a further seven in 216 BC. This increased further to twenty legions by 203 BC, reduced to sixteen legions by the end of the war in 201 BC, and further reduced to six legions for peacetime (Keppie, 1984, 97–100).

By the late Republican period, legions were using identifying numbers ranging from I to XVIII. Of these, numbers I to IV were still allocated to the consuls. However, with the exception of these consular legions, a system of clockwise numbering appears to have been used for the remaining legions, V for the legion in Spain, VII, VIII, IX, X in Transalpine and Cisalpine Gaul, with higher numbers for Macedonia and the East, and XVIII in Cilicia, in 56–54 BC (Keppie, 1984, 97–100).

TRANSITION TO EMPIRE – CHANGES TO THE LEGIONS UNDER THE 'TRIUMVIRATES'

Under pressure to reward time-served army veterans with land, the 'First Triumvirate' of Pompey the Great, Crassus and Julius Caesar, formed in 59 BC, led to a restructuring of the legions and a further increase in their numbers (Keppie, 1984, 71; Broadhead, 2007, 157–160).

With Crassus campaigning in Parthia, Pompey in Spain, and Caesar in Gaul, existing legions were divided between them, the identification numbering becoming more complicated following Caesar's Gallic Wars and subsequent Civil Wars. In 58 BC, as proconsul in Gaul, Caesar had four legions under his command (VII to X). To these he added a further four (XI to XIV) to fight the *Helvetii* and the *Belgae* of north-eastern Gaul, adding another (XV) in 53 BC, leaving him in control of nine legions (Keppie, 1984, 97–100).

After the death of Crassus in 52 BC, a breakdown between Pompey and Caesar led to the 'Civil' Wars from 50 BC, when Caesar famously crossed the Rubicon with *legio* XIII, increasing his army to twelve legions the following year (Keppie, 1984, 97–100). Caesar then continued to raise more legions, adding numbers up to XXX (the latter at Mutina), including his famous *legio* V *Alaudae*, the 'Larks', levied in Cisalpine Gaul. By the Battle of Pharsalus in 48 BC, Caesar had raised yet more legions (including four new numbers, I to IV, as these were traditionally commanded by the consul), bringing the full sequence (I to XXXIII) under his command. After the battle, further regrouping of Pompey's surviving troops added numbers XXXIV to XXXVII (Keppie, 1984, 103–106).

Between 47 and 44 BC, legions VI to XIV were retired from service as time-served (six years' service being expected at that time), but after Caesar's death in 44 BC some of these were reformed. On Caesar's death, Rome's (and Caesar's) wealth was controlled by Mark Antony. However, Caesar had made his nephew, G. Octavius ('Octavian', born 63 BC), his protégé and his heir. Initially the two rivals campaigned against each other, but they then joined forces against Caesar's assassins, forming a second triumvirate with Lepidus (Lepidus in the east, Octavian in the west, and Antony in the south/Egypt), Antony forming an alliance with Cleopatra during his campaigns to Egypt. Keppie (1984, 132–135) reports that, to provide manpower for these campaigns, Octavian reformed Caesar's 'retired' legions, VII and VIII, in Campania; V *Alaudae* was regrouped by Antony; and legions VI and X were re-established by Lepidus. The fate of the remaining disbanded legions (from VI to XIV) remains hazy. Keppie (*ibid.*) suggested that some may have been with Antony (including XII 'Antiqua'), but others may not have been reformed at all (e.g. XI, XII and XIV).

As with the previous Triumvirate, however, relations soon broke down, Antony depriving Lepidus of his provinces (accusing him of collusion with an enemy, Sextus Pompey). Despite divisions between Antony and Octavian, the remaining legions were not immediately renumbered after the Battle of Philippi (42 BC). However, many time-served men were later released, and while Antony retained some of Caesar's old legions, including V *Alaudae*, VI *Ferrata*, X *Equestris*, XII *Fulminata* and III *Gallica*, Octavian regrouped the remainder into eleven legions, adding further legions between 42 and 32 BC to fill out numbers. Some legion numbers therefore came to be duplicated with those commanded by Antony (new V *Macedonica*, VI *Victrix* and X *Fretensis*; Keppie, 1984, 132–135).

After Actium, Octavius (now renamed 'Augustus') then 'retired' 100,000 men from the disbanded legions (who were all set up with gratuities and land provided by the state, in veteran colonies either in Italy or in the provinces), adding the Antonian legions to his own, keeping their existing numbers (with duplications). These were then identified by 'names', some legions being given the name '*Gemina*' (twin), for example the X *Gemina*. Retaining eighteen legions, he then added eight more in AD 6, and another two in AD 9, leaving a total of twenty-eight legions, although legions XVII to XIX were lost on the lower Rhine in AD 9 and were not replaced, these being the three 'lost' legions of Varus (Keppie, 1984, 132–135).

THE AUXILIA

Following Caesar's model, Augustus formed more permanent auxiliary units of specialist troops to support the legions. These comprised cavalry, archers, slingers and specialist light infantry troops. The majority of these units were formed up under the command of their own chieftains. They wore native dress, used their own indigenous equipment and fought using their own specialist weaponry.

However, following a number of rebellions by auxiliary troops – including the Pannonian Revolt, AD 6–9; the German revolt of the *Cherusci*, under the command of Arminius, AD 9; the Thracian revolt, AD 46 (Luttwark, 1999, 13–20); and the rebellion of Batavians under the command of Julius Civilis, AD 69 (Cary & Scullard, 1979, 418–19, 509) – in particular that of Julius Civilis' Batavians, all auxiliary units were disbanded shortly after the time of Nero. They were then reformed, no longer under the command of their own chieftains, but to be commanded by a Roman tribune, aided by two prefects and equipped with Roman-style equipment (helmets and mail or scale cuirasses). The specialist troops (such as the archers) retained their own native weaponry (where it was without Roman equivalent), but the infantry were equipped with a Roman-style short sword and spear (*gladius* and *hasta*), and the cavalry equipped with Roman-style lance or spear (*hasta*), along with either the long cavalry sword (*spatha*), or the alternate 'Spanish' sword, the *falcatta* (Webster, 1998, 141–155).

The overall identity of these troops also changed over time by the second century AD. Initially these had been non-Roman provincial troops of distinct ethnic origins. However, with units based away from their homeland, and fresh recruitment to the units made in other provinces, their ethnic identity became diluted. Some individuals were given citizenship on retirement (their sons then joining the units later with citizen status). There were also incidents of citizens choosing to join these units to serve as auxiliaries, rather than as legionaries. Furthermore, some whole units had been granted citizenship en masse as recognition of actions on campaign, as for example with the *cohors I Lepidiana equitata civium Romanorum* (Gilliver, 2007, 194).

The changes to the regular auxiliary units also forced the Roman army to introduce different types of auxiliary support, which Gilliver (2007, 195) described as the 'irregular' ethnic units of '*numeri*' and '*nationes*' (the most famous of these being the Palmyrian archers, used for the first time in Judea in AD 70–71; Southern, 1988, 89–92). These again were native troops, fighting in their own clothing, with their own specialist weapons, their lower commanding officers drawn from their own chieftains, but with an overall commander who was Roman. They differed from the standard auxiliary units in that they were units of non-Romans, often supplied by 'client' leaders, whose service was 'given' for short periods to an individual, their 'patron'. Their loyalty was therefore not primarily to Rome, but to this individual 'patron' commanding the army.

Class	Property (asses)	Equipment	Centuries		
			Juniores	Seniores	Total
I	100,000	Helmet, round shield, greaves, cuirass, spear, sword	40	40	80
II	75,000	Helmet, oblong shield, greaves, spear, sword	10	10	20
III	50,000	Helmet, oblong shield, spear, sword	10	10	20
IV	25,000	(Oblong shield in Livy), spear, javelin	10	10	20
V	11,000	Sling, stones, (javelin)	15	15	30
				Infantry Total	170

Table 9. The infantry in the Servian system (after Goldsworthy, 2003).

APPENDIX II

EQUIPMENT & *FABRICAE*

THE EVOLUTION OF THE EQUIPMENT PANOPLY

There are numerous examples throughout the history of the Roman army whereby the valued qualities of an enemy's equipment and tactics were observed, then assimilated into their own, in order to use these against them. Rawlings (2007, 54), however, considered this to be part of a more complex process of military interaction and interchange throughout Italy and the Mediterranean, as evidenced in tomb paintings and pottery decoration.

In the fourth century BC the most common military equipment used was the 'hoplite panoply', although with some regional variation, adapted to local conditions and preferences. These were not exclusively used in hoplite battle formation, but were combined with the use of large thrusting/stabbing blades and light throwing javelins, which may indicate the use of more open formations (Rawlings, 2007, 45–62). By the fourth and third centuries BC some experimental use of '*proto-pila*' has been proposed, although Rawlings (*ibid.*) acknowledged the possibility that the standard second-century BC *pilum* may have developed from observation of equipment used by third-century BC Spanish and Celtiberian Carthaginian mercenaries.

It may have also been about this time that the round shield was phased out in favour of the long, oval *scutum* (as added protection for those soldiers equipped with less armour), with lighter javelins replacing the heavy lance (Feugère, 2002, 38). Diodorus (23.2) also considered the principle change of this period to be the introduction of the oval shield (*scutum*), which had been popular with the Gauls of northern Italy and is also seen on southern Italian vase and tomb paintings as an alternative to the circular hoplite shield.

Diodorus (23.2) also attributed at this same time a change from phalanx to *manipular* formation, although modern sources disagree on when this change was made. Rawlings (2007, 55), for example, considered the transition to be gradual, with the *triarii* of the second century BC still operating in a similar way to the hoplite phalanx. He also observed that Hellenistic armies of the later fourth century BC were beginning to use oval shields (known as *thyreos*), with groups of infantry called '*thyreaphoroi*' described in the third century by Plutarch (*Philip*, 9.1–2), who told of third-century BC Achaeans using *thyreos* and light spears. However, these were easily forced back and scattered in close-quarters melees, apparently because they still used the new equipment with the old phalanx formation, rather than in more flexible maniples.

In contrast, Rawlings (2007, 57) saw the Roman soldier's oath (*coniuriato*) not to flee or break ranks as indicative of their use of more flexible battle formation, rather than the rigid phalanx battle order, citing Livy (22.38.2–5), who noted the exemption for soldiers to leave their lines in order 'to recover or fetch a weapon, save a friend or strike an enemy'. This would then permit troops fighting in small units to break lines periodically to collect scattered javelins for reuse, as described by Livy (10.29.6) at the Battle of Sentium against the Gauls (Rawlings, 2007, 57).

Livy (8.9–13) also described the use of *manipular* formation in 338 BC, although this is not the same as Polybius' description from the second century BC, but indicates that only the third line of troops (*triarii*) used javelins (*hastae*). Hoyos (2007, 69) interpreted this to suggest that the remaining two lines of heavy infantry must therefore have used *pilum* and *gladius*. Both Livy (31.34.4) and Polybius (6.23.6) described a 'Spanish' short sword which may have been the *gladius*, noting that both pieces of equipment and *manipular* array were in standard use by the time of Polybius' accounts of battles against the Gauls in 225 and 223 BC (2.30.8, 33.4.6).

Polybius further described the differences of equipment used at that time by the different ranks within the battle formation, denoting the different functions served. For example, the *velites* were equipped with a sword, a javelin and a small round shield (*parma*), better suited to light skirmish actions. The *hastatae* and *principes*, in contrast, used the oval *scutum* and *gladius*, along with two *pila* (one heavy and one light). Keppie (1984, 35) considered the *gladius* may have been adopted from Spanish auxiliaries serving with Hannibal, who, conversely, had also adopted Roman equipment and assimilated Roman fighting techniques for his own troops.

Less wealthy soldiers at this time would have worn a simple plate strapped to the chest, although the older, more experienced and wealthier men would have worn more substantial protection, possibly by this time a mail cuirass, or *lorica hamata* (as described by Polybius; 6.23). This could perhaps have been combined with shoulder doubling for added protection, as seen on the Altar of Ahenobarbus (Fig. 13), although this example dates to the first century BC. This tradition appears to have been adopted from Gallic tribes of northern Italy, along with the tradition for Montefortino-type helmets (Burns, 2003, 71).

In contrast to the Roman military system, the opposing Carthaginians at this time did not use armies of temporary citizen militia. Their army was made up of contingents who owed loyalty to a particular commander, but who did not always co-operate readily with different leaders, leading on occasions to 'infighting' according to Polybius. They were professional soldiers of varying nationalities, hired as mercenaries and originating from Africa and parts of Europe, including Libyan heavy cavalry, Numidian light cavalry and infantry from Spain and the Gallic tribes of northern Italy. Their equipment would in many respects be similar to that of the Romans, the mail cuirasses appearing to be of Gallic origin, as are the Montefortino helmets. Both armour types originated from regions known to have supplied Carthaginian mercenaries, as well as their equipment, as suggested by finds of Montefortino-type helmets from Villaricos, in Almeria, which was reputedly an important recruiting centre for Punic generals (Quesada Sanz, 1997, 155).

Polybius, in describing the recruitment process of the Roman army, related that all eligible men of military age (seventeen to forty-six years) were normally called to the Capitol in Rome, and from these men the best candidates (*dilectus*) were selected. These recruits were then dismissed to arm themselves and regroup elsewhere. Paddock suggested that the recruits would either equip themselves prior to departure, or close to this assembly point, possibly being buried with these arms after their eventual return home years later. He cited as evidence helmets found from fourth- and third-century BC grave contexts in Perugia, Etruria, which he believed to be the location of a manufacturing centre (Paddock, 1985, 143).

With the change to the cohortal system, the spear fell out of general use (as less suited to the cohort formation), being replaced with a more standardised weapon assemblage of javelin (*pilum*), short sword (*gladius hispaniensis*) and dagger (*pugio*):

- The *pilum* was designed to penetrate an opponent's shield, then break or bend so that it could not be thrown back.
- The *gladius* was a short (30 inch or 75 cm) slashing sword, with its weight balanced one-third from its tip.
- The *pugio* was a short dagger, carried on the belt, which appears to have served a more utilitarian purpose than for combat (in re-enactment contexts being found to present more danger to the bearer than an opponent, from its tendency to 'accidentally' fall out with alarming regularity, with the result that many 'owners' have them permanently fixed into their sheaths).
- The *casis* (helmet) provided unobstructed vision and hearing, permitting the wearer to see or hear orders (by *signa* or *cornu* respectively).

- The *scutum* was a curved oval shield formed from two or three layers of plywood, with covering of calfskin, 4 feet by 2.5 feet (120 cm by 75 cm) as an average size, with bronze or iron edge binding, central spine (*spina*), and an iron or copper alloy boss (*umbo*).

Polybius (6.39.15) related that it became necessary for the state to buy equipment for the poorer *assidui*, who were now admitted following further relaxations of the property requirements. The soldiers did not receive this equipment free of charge, however, being required to reimburse the state for its cost in instalments deducted from their *stipendium*. In 123 BC, Gaius Gracchus tried to make it illegal to make deductions for food, clothes or equipment (Plutarch, *C. Gracchus*, 5.1), but was unsuccessful and the deductions continued, as described by Tacitus in AD 14 (*Ann.*, 1.17.6). Livy related how this equipment was originally sourced from private manufacturers (22.57.11), with a further reference provided by Cicero (*Rab. Perd.* 20) to state-owned armouries and arsenals in 100 BC.

Marius instituted an overall revision of military equipment, standardising and improving armour and weapons, ensuring that all men received mail armour, a standard sword and long javelin (*pilum*). He is also credited with the redesign of the *pilum*. Unchanged for centuries, it was now designed to bend on impact to prevent reuse by an enemy. Marius designed a wooden peg to secure the head to the shaft, which would then snap on impact. The *pilum* could still not be reused by an enemy during the course of a battle, but could later be retrieved after the conflict and quickly repaired by replacement of the wooden peg (Scullard, 1970, 58–59; Campbell, 2000, 6).

Although Marius' reforms may have had a positive effect on the organisation and efficiency of the army, the same cannot be said of the quality of its equipment. While the individual soldier had provided his own equipment, and while membership of the army was restricted to the more affluent members of society, the quality of their equipment remained high, in that they could afford the work of trained craftsmen, producing individual pieces to order. With the admittance of the poorer classes, who had to be equipped by the state, along with the necessity to have to kit out large numbers of men at short notice, quality suffered with the economies of bulk buying. This is particularly noticeable in the helmets produced at this time, with the known examples crudely formed, with minimal decoration, and even on occasions not fully finished off, or left rough from the raising process, not all of the surface hammer marks being planished off (Paddock, 1985, 144).

This decline in quality became more marked in the ensuing Social War of 91–87 BC, and the subsequent civil wars, the former event providing more potential recruits, with the relaxation of citizenship to include those from all neighbouring Italian states south of the River Po (Campbell, 2000, 7).

In 49 BC, Caesar increased his army to twelve legions. Where previously soldiers had been paid amounts barely sufficient to cover the cost of food, he doubled their pay, encouraging a greater volume of volunteer 'professional' soldiers from the lower social classes. With the death of Caesar in 44 BC, until Octavian's defeat of Mark Antony in 31 BC, military growth slowed. Existing equipment appears to have been stockpiled or remained in service for long periods, and demand fell for new equipment with little evidence for the recycling or manufacture of new goods, and with few 'innovations' introduced to the equipment assemblage. The 'retiring' 100,000 men from the disbanded legions were all set up with gratuities and land provided by the state, in veteran colonies either in Italy or in the provinces, setting a precedent for army retirement policies (Scullard, 1970, 216).

The next major reorganisation of the army by Augustus, with his reforms of 27 BC, finally changed the 'Republican' army into the 'Imperial' army. Membership of the army was no longer seen as a disruption, but became a career of choice, with citizens serving a minimum of sixteen years. The primary purpose of the army had changed, however, no longer being the great wars of conquest, as during the Republic, but as low-profit police actions in defence of the Empire. When production recommenced, evidence of new methods of mass production began to appear, such as spinning of helmets (Paddock, 1985, 148).

With the Augustan reforms, the Roman army became a true 'professional' army for the new Empire, and any future changes to equipment and fighting methods were event driven in response to changing enemies and their chosen weapon types.

THE MANUFACTURE AND THE PRODUCTION OF ROMAN MILITARY EQUIPMENT FROM PRIVATE COMMISSIONS TO STATE ARMOURIES (*FABRICAE*)

Up until the later stages of the Republican period armour was commissioned by the individual. This was then sourced from independent craftsmen, who operated from small-scale workshops, following their own traditional designs. As the boundaries of property ownership relaxed and less wealthy individuals increasingly participated in military duties, they were unable to meet the full cost of armour and weaponry provision personally (Polybius, 6.39.15). The state was therefore required initially to meet that expense, although recouping some of the cost later through deductions from the *stipendium* (Livy, 22.57.11, relating that equipment was originally sourced from private manufacturers). It would therefore seem reasonable that in order to kit out large numbers of these poorer *assidui*, the state would commission equipment in larger quantities than could be supplied by the smaller independent workshops, which may then have required the establishment of larger dedicated arms factories (*fabricae*), with Cicero (*Rab. Perd.* 20) referring to state-owned armouries and arsenals in 100 BC.

The location of these large arms factories and how they were supplied, manned and operated has been the subject of much discussion (with particular recognition given to the work of Bishop, 1985, for the early period of the Principate, and to James, 1988, for the later Empire). No precise locations have been absolutely identified, although some speculative sites have been proposed based on building types and proximity to artefacts such as tools and waste materials, as well as the limited epigraphic, literary and sub-literary evidence for the later periods.

Bishop (1985) described his perception of the 'legionary' *fabricae* of the Principate of the early first century AD. These he saw as firmly controlled by the military (as opposed to the later *fabricae*, controlled by the state, as described by James, 1988, 257–331). Although describing the organisational and command structures of the first century AD, Bishop uses as his source Vegetius' *De Rei Militari* (dating from the fourth century AD) and a fragment by Tarruntenus Paternus in Justinian's *Digest* (from the end of the second century AD). Vegetius, however, had made use of earlier sources, including Iulius Frontinus and Cornelius Celsus (both from the first century AD), and the latter of these used Cato (dating from the Republican period). Their views on the organisational and command structures may therefore at times be inaccurate, reflecting earlier or later periods. Vegetius related that the officer in overall charge would have held the position of *praefectus castrorum* (or *quaestor* in the Republican period), with the individual workshops controlled by an *optio fabricae*.

Bishop also made use of the sub-literary sources of writing tablets from Vindolanda, and a papyrus from Berlin, which describes the numbers of men employed in a 'legionary' *fabrica* on a single day (100 at the former and 340 at the latter), with a breakdown of their status. This would appear to have consisted of a core group of skilled workers and specialist craftsmen (*immunes*), with unskilled labour supplied by the soldiers (*cohortales*), associated civilians and a group of possible slaves. Both of these *fabricae* appear to have been involved in general armour and weapon production: the German site producing *spathae*, two types of shield, iron plates (possibly for use in segmented armour or helmets), bows and catapult fittings; whereas at Vindolanda, mention is made of *gladiarii*, *fabri* and *scutarii* (Bishop, 1985, 3).

Although there are no positively identified *fabricae*, Bishop (1985, 7), cites the possible first- and second-century AD locations of Haltern, Hofheim, Valkenburg, Oberstimm, Inchtuthil, Wiesbaden and Vindolanda, where similarities of buildings (of a series of small rooms, arranged around a central courtyard), have been seen, often with an associated water supply and occasionally with proximity to evidence of hearths. In some instances these distinctive building types have been identified within a fort itself, but on other occasions they appear within dedicated walled 'industrial' enclosures, often annexed to a fort (or within an otherwise 'civilian' town, such as at Corbridge, and as more recently found on the outskirts of the settlement at Vindolanda, presumably where industrial activity could be kept under direct military supervision.

Further secondary evidence of manufacture/repair workshops may come from waste materials (metal, crucibles, wood, bone or leather), or possible evidence of recycling, as in the case of a quantity of nails and iron tyres from Inchtuthil. Bishop (1985, 9) suggested that many of these *fabricae* may have practised a high level of repair/recycling in addition to manufacture, repairing damaged items where possible and recycling at the end of their usefulness, in order to keep the need for new raw materials to a minimum, citing Vegetius' ideal of military self-sufficiency. Bishop proposed that the large quantities of materials deposited in pits and ditches, on abandonment and demolition of sites such as Newstead, are indicative of deliberate retention of scrap for potential reuse, rather than accidental loss (*ibid.*).

Bishop also suggested that objects may remain in use for long periods (as evidenced by signs of wear and combat damage, subsequent repairs and inscribed names of multiple owners), offering a notional 'twenty man-years period' for the lifetime of a piece of equipment in times of peace. However, it should be remembered that this 'notional' lifetime would be greatly reduced under combat situations, and would also be relative to the type of equipment under discussion. For example, more resilient parts of the panoply (such as helmets, plate armour, etc.), would need less repair and remain in use through more owners and for longer periods (possibly considerably more than this 'notional' twenty years), than would items made of more perishable materials (such as shoes or shields, made from leather and wood).

Bishop (1985, 15) also considered the possible regional differences in armour/equipment production across the Empire, proposing that distinctive features could be identified in the work of individual craftsmen operating with specific legions (as seen, for example, in the 'Upper German' style of decoration of belt plates and scabbard fittings dating from the Tiberio-Claudian period), although some exchange of ideas may have occurred between legions during joint campaigns. He also proposed that although the 'legionary' *fabrica* system operated in the west and along the frontier zones, a different system (the Greek-style '*polis*' system) could be seen in the Mediterranean zone and in the east, whereby individual city states were able to organise their citizenry to produce large quantities of equipment at short notice, in order to rapidly equip large armies. This 'mass production' he believed to be responsible for the poorer quality of the Imperial-Italic helmets produced by these workshops (Bishop, 1985, 16).

During the centuries immediately following, the organisation of arms and armour production is unclear, although it would appear that, as frontiers became less volatile and as there was a movement from armies of campaign towards standing armies, an increasing level of manufacture may have passed back to the private sector. This may have commenced with the provision of short-life equipment, such as leather goods, but later on may be reflected in other 'heavier' industries, such as iron production and metalworking. For example, there is a suggestion from manufacturing sites such as Templeborough of an abandoning in later periods of the military presence (or a reduction to a minimum garrison staffing), with an apparent 'civilian' continuance of industrial activity moving back inside the fort itself.

Towards the later stages of the Empire, the production of military equipment took a cyclical change, away from private enterprise, to a controlled, organised system, this time under direct state control rather than by the military. This transition was discussed at length by James (1988, 259–331). In his discussion of the arms manufacturing sites of the later Empire, James relied mostly on the names of state *fabricae* listed in the *Notitia Dignitatum* (along with other military and government installations under the control of their relevant *Magistri*). Dating to the fifth century, this appears in two versions: the *Notitia Dignitatum Occidentalis*, dealing with the western Empire, and the *Notitia Dignitatum*

Orientalis, dealing with the east. James identified an organised network of *fabricae*, twenty in the western Empire, fifteen in the east (along the Rhine and Danube frontiers, and evenly spread around the eastern provinces, although with some areas, notably Britain, Egypt and the south-western provinces, where no *fabricae* appear to exist), which he categorised into those for general armour manufacture and those for the manufacture of specific/specialised equipment (1988, 261).

These categories of *fabricae*, according to James (1988, 261), are:

Specialist Equipment
Arcuaria – bows; *Sagittaria* – arrows; *Ballistaria* – artillery; *Hastaria* – spears; *Spatharia* – swords; *Scutaria* – shields; *Armamenta* – marine fittings, but could mean arsenal, armoury or weapon store.

General Equipment (armorum *&* clibanaria *&* loricaria)
Armorum – James suggested these as general workshops for any part of the panoply (helmet, body armour, weapons), except missiles, and possibly shields (as references are also found to '*scutaria et armorum*', where shields were made at separate locations). Also possibly used for sword production in the east, where there are no *fabricae spathariae* identified.
Clibanaria – mostly found in the east (where most of *catafractarii* and *clibanarii* were located), which James interpreted to suggest heavy cavalry armour production.
Loricaria – only in found in the west. James did not support the suggestion that these are for leather defences, but considered it a western term for body armour, with swords coming from *fabricae spathariae*.

Those which he categorised as for general production, he identified as being located in pairs in each of the western dioceses, along with a single *scutaria* for shield production, although with no such pattern visible in the east, where there was no linear frontier. The more specialised *fabricae* then appear to have been irregularly placed, dependant on the needs of the relevant specialist units. For example, missile and archery equipment production was all located in the west. This James related to archery being an eastern tradition, where archers would have made their own bows and ammunitions, with additional equipment possibly produced as a 'tax in kind' for export to other areas. In the west, however, where archery was not a local tradition, these had to be supplied, with the *fabrica* at Ticinum being centrally located to best import component parts and to distribute the end product (James, 1988, 264).

James discussed the locating of these *fabricae* in strategic positions along the frontiers, and in an organised distribution in the eastern provinces, contrasting their absence in regions such as Britain and Egypt, which were no longer by that time of comparable military significance. He suggests that the majority of these *fabricae* are a phenomenon of Diocletian's building programme and reorganisation of the provinces. The army had become less self-sufficient after a period of civil wars, foreign invasions and political upheavals, making it more dependent on civilian workshops. Inflation of the third century AD had collapsed the coinage (and along with it, the private market), leading to an increased use of 'taxes in kind', with the almost exclusive outlet for civilian armourers' goods now being the army. A state 'nationalisation' of armour production then became the next logical step, with its manpower (*fabricenses*) comprised of well-paid freemen who were tied for life to their trade by legislation (Theodosius II, AD 438, *Novellae*, 6), with sons following their fathers into their profession, but with additional recruitment from veterans and some volunteers. These artisans would have held a similar status to the militia, with the same privileges, exemptions and rights to draw financial support from the military tax, the *annonia militaris*, organised with a military-style command structure into a corporate body similar to the civilian guilds. James (1988, 276) then used his list of thirty-five known *fabricae* of the fifth century AD, compiled from the combined western and eastern editions of the *Notitia Dignitatum*, along with an estimation of 400 to 500 men being active at each site, to propose a total of 7,000 to 17,500 men being involved in armour production at that time.

James acknowledged that not all *fabricae* were established at this time, but claimed that this was the time when the organised network of paired *fabricae* appears to have been initiated, and that many may have preceded this time, some possibly originating from earlier legionary-based *fabricae* (although this he viewed as 'nominal'). This then suggests a differentiation between the early and later *fabricae*, as two separate entities, albeit that the latter may have derived from the former, which James saw as the 'legionary' and the 'imperial' respectively. He cites the reduced quality of equipment at that time, and the abrupt change to the simplified, functional, bipartite and multi-pieced helmets, with armourers being directed to work to government quotas of a specific quantity and quality at a minimum cost (1988, 271).

These later *fabricae*, by the third century AD, were no longer based at 'military' locations, but on 'urban' sites where major cities could supply a ready workforce, ideally in areas where communities of armourers were traditionally found (often in mining areas, with advanced pre-Roman metalworking and arms production traditions). While proximity to raw materials was an important factor in location, James (1988, 267) considered that this was only contributory to the other factors of manpower, personal supply and distribution: where 'taxes in kind' could provide food, materials and bullion, and the roads and navigable waterways network provided transport for raw materials and distribution of finished products.

Although the *fabricae* of this period were on the periphery of the armour manufacture discussed in this study, evidence for their organisational structure, location and manpower are of relevance to the earlier periods as they show a developmental progression, in both the arms industry itself and in the equipment produced by it, helping to explain some of the subsequent changes in quality, design and ethnic influences seen on the artefactual remains. The lists also demonstrate the use of general-function and specialist *fabricae*, the latter dedicated to the production of different types of equipment, linked to distinct units of specialist troops, which suggests that a similar segregation of production may have been the case in earlier periods, which may aid with the identification of such earlier-period sites in the future.

Developmental changes can therefore be seen as production moving cyclically, from small-scale private civilian workshops to an organised, large-scale military operation, with a gradual gravitation back towards private enterprise, and once more into a planned, large-scale, state-controlled operation. As the 'control' and level of production fluctuated, so did the level of quality and workmanship of the finished products, which implies different methods of manufacture, and this may then be evident in the archaeological record (in building design, site location and layout, tools used and waste materials produced). This also may have implications for the consumption of raw materials, the rate of repair/replacement, and the use to which the equipment is put (with considerations such as the resilience of the panoply in comparison against lightness of use and ease/cost of replacement as a short-life commodity).

WOOD

EVIDENCE FOR WOODWORKING

At first glance it may not seem obvious how wood features in the manufacture of military equipment, and indeed it is not a component of wearable equipment. However, it is used for construction of shield boards, shafts for spears and javelins (*pila* and *hasta*), in addition to peripheral use as fuel or as construction material for workshop buildings (*fabricae*). For each of these uses, different wood types would have been selected based on the suitability of their physical properties, tempered, of course, against availability.

For example, both Polybius and Pliny advised on the use of wood in shield construction, Polybius describing the method of producing ply (Polybius, 6.23.2–7) and Pliny describing the best timbers to use, specifically that

> the most flexible and consequently the most suitable for making shields, are those in which an incision draws together at once and closes up its own wound, and which consequently is the more obstinate in allowing steel to penetrate; this class contains the vine, *agnus castus*, willow, lime, birch, elder and both kinds of poplar. Of these woods the lightest and consequently the most useful are the *agnus castus* and willow ... Plane has flexibility, but of a moist kind, like alder; a drier flexibility belongs to elm, ash, mulberry and cherry, but it is heavier (Pliny, *Hist. Nat.* 16.77; James, 2004, 167).

The choice of wood type would therefore be partly dependant on what timber was readily available (either locally, or within reasonable distance for importation), and on its intended purpose. For shields it would be preferable to have flexibility, particularly in the outer plyboard layers, with a strong, hard wood such as oak in the centre. However, the properties of these different wood types would be further influenced by the way in which the timber was treated, in 'seasoning', and in the way that the plank is cut from the trunk. Living wood is composed of cellulose-walled cells, containing living matter in the centre. Trees grow by building moist, flexible layers of these cells (sapwood) at their outer edge, in the Cambium layer, under the bark. Towards the centre, the cellulose in the older cells is replaced with a material called lignin, and the living central part dies, forming the drier, harder, but less flexible heartwood (Hodges, 1976, 113). This heartwood would then have been more suitable, because of its strength and rigidity, for middle ply layers, the more flexible sapwood being preferable for outer ply, where flexibility was required.

The Roman author Vegetius tells how the military attempted to make each fort as self-sufficient as possible, making it their chief care to have everything for the service of the army within the camp, and this would have no doubt also extended to a reasonably close distance around the fort (Vegetius, *Epitoma rei militaris*, 2.10). However, any description of military methods and events by Vegetius must be read with caution. He was an author writing in the later periods, casting his mind back to earlier times of Rome's 'glory'. Rather than describing actual practices, often he would make suggestions to methods of 'best practice' based on his perceived view of the past. Therefore, if he stated that all forts were self-sufficient, this may be 'wishful thinking', the reality being less absolute, although the probability is that they had always striven towards self-sufficiency since the decision to withdraw support to overseas troops at the time of Fabius Maximus (although 'self-sufficiency' at that time comprised a reliance on taking 'booty'; Cary & Scullard, 1979, 130).

Despite this, on the initial foundation of a new fort, perhaps where the Roman forces were first establishing a base in an area, for reasons of security it would have been logical to import wood from more secure regions, using the road and river infrastructure. In more peaceful times, once the fort was established and the surrounding area (or *territoria*) securely under Roman control, it is probable that, in locations where trees were readily available, timber would have been sourced locally. Wherever possible, those tree types best suited to specific needs (although not necessarily the absolute ideal) would have been selected from local resources; for example, while the shield from Fayum was made from strips of birch, those from Doncaster and Dura Europos were made of oak/alder and plane/poplar respectively (Kimmig, 1940, 106–111; Biek, 1978, 269; James, 2004, 162–163).

Where wooden artefacts are found (shields, weapon shafts, etc.), their wood type does not necessarily reflect the material locally available, as these objects are, in their very nature, personal and portable, and their owners could have travelled considerable distances on campaign before discarding them. However, as these are also often short-life items of equipment, easily damaged or intended for 'throw-away' use (quite literally, in the case of *pila* and *hasta*), it is probable that many similar 'consumable' items would constantly need to be replaced in the military workshops, so a high percentage of those in use at any time are likely to have been manufactured from local materials.

The earliest forts, on initial foundation, would have been built from timber, only being rebuilt in stone in their later phases. Standardisation of fort design, however, suggests that many of the component parts would have been prefabricated and are likely to have been

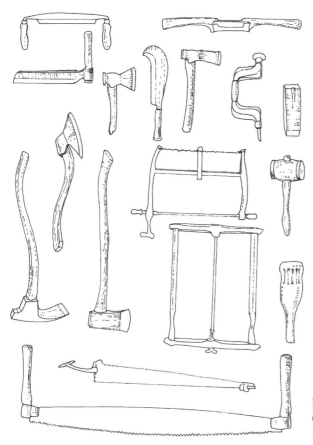

Fig 77. Traditional wood tools.
(Artwork by J. R. Travis, from Sparkes, 1997, p. 16)

brought in from outside the area, some buildings possibly even having been dismantled from their original locations and reassembled at the new fort. For new-build projects, possibly in less-than-friendly environments, it would have been a logical choice to have as many of the requirements of a new installation prefabricated, ready for rapid assembly. Wood was also used frequently as a building material, even into the later phases of a fort's development where there was a substantial level of stone construction, particularly for lower-status workshop buildings of a temporary nature, or for storage buildings. Evidence from recent excavations at Carlisle suggests that wooden buildings may have been periodically burned, in order to clear the site of any parasitic infestations (J. Zant, pers. comm., October 2005). In the case of some buildings of intended short life, the timber used in the construction had not been stripped of bark, although the opposite is more usual, as timber with bark remaining will generally rot much quicker than that where it has been removed, the removed bark being useful in other industrial processes, such as tanning of leather.

Wood types selected by preference for building purposes may not necessarily have matched those used for weapon shafts or shields, but again, timber may have been sourced locally (in non-hostile locations, where safe to do so). Preliminary treatment of the logs would then have been carried out close to the source ('thinning out' – trimming of excess foliage, etc.), prior to any secondary processing ('barking', seasoning, splitting, sawing, shaping, etc.) or removal to the fort. Even with locally sourced timber, this primary location would not have been immediately adjacent to the fort (as the immediate area around a fort would have been cleared of trees and undergrowth for reasons of security), and these activities would be unlikely to leave easily identifiable archaeological traces (the product being natural and biodegradable).

PROCESSES

Prior to the mid-twentieth century, techniques and tools used for processing timber had changed little from the Roman period (with some exceptions in the seasoning and division of the trunks) and, with the exception of modern mass-processing methods, analogies can be drawn from studies of the traditional toolsets and methods of woodland crafts still in small-scale operation today.

Felling
The first stage in any use of wood has to be the felling of the timber itself. Where possible, this would have been carried out in the autumn and winter periods when the tree would be at its most dormant, when less sappy, as this would produce timber which was of a better colour and more durable. In the absence of modern large machinery, this would have been mainly achieved with an axe (by preference rather than by use of a saw, as the sap within the living tree would tend to cause the wood to stick to the saw's blade). Up to the nineteenth century, the typical felling axe would have had a wedge-shaped, straight-edged blade set on a long, straight handle (Sparkes, 1991, 3).

After removing any swellings at the base of the trunk (traditionally known as 'laying up', 'rounding up', 'facing up' or 'setting up'), an 'undercutting' would be made on the front face (known as 'birdsmont', 'sink' or 'kerf'). Then a 'crosscut' saw would be used to cut into the trunk from the back (with wedges inserted as temporary supports during this process), then, once the 'crosscut' was complete, the tree would be pushed forward towards the 'undercutting'. Once felled, the trunk would be 'thinned out' – trimmed of side branches ('top and lop') using a smaller, angled 'siding-up' axe. These surplus branches would not be entirely discarded, however, as they could be used for brushwood, for fuel, or for production of charcoal (Hodges, 1976, 113; Sparkes, 1991, 3).

Barking
Although there is evidence of some buildings constructed of un-barked timber at Vindolanda, this would have been more prone to rotting and would only have been used for hurried,

short-life constructions. Most trunks would then be stripped of bark before any further splitting, cutting or shaping, probably before any removal from the original source to any secondary processing location. Bark could even be removed prior to felling of the tree, although if left until spring (April to June), removal is easier. The bark by-product, once removed, being high in tannic acid, would then have been reused in other industrial processes, such as tanning/curing of leather (oak bark being best for this purpose, but alder, beech and willow also being used; Sparkes, 1991, 4).

The bark would be peeled off in bands, cutting an upper and lower ring around the trunk circumference to form a wide strip using a hatchet, then making a vertical cut between these rings with a 'barking iron' (a flat, round, bladed chisel with a sharpened edge). The iron was then inserted into this cut and used to lever off the bark in a wide strip, which was then dried on racks or poles. As part of the curing process for leather, this dried bark would then be soaked in water in pits to extract the tannin, and skins soaked for several months in the mixture (Sparkes, 1991, 4).

Seasoning

Living trees contain a large amount of moisture, in the form of sap, in varying amounts throughout the trunk (mostly in the moister 'sapwood' at the outer edge, with drier 'heartwood' towards the centre). Once the tree is felled, however, the sap will cease to flow and will gradually dry out over time, affecting its dimensions and strength and flexibility. Therefore, before the felled timber can be used, the wood needs to be 'seasoned', removing this excess moisture in a controlled manner in order to minimise shrinkage, distortion and splitting. In the seasoning process, marked differences can be seen between modern mass-production methods and processes used in antiquity (and down to even just the last century).

In modern practice, for reasons of speedy processing, sacrificing quality for quantity of production, trunks are rough sawn into planks soon after felling. The planks can then be artificially dried in ovens, their lesser thickness allowing them to dry out more rapidly than would be the case for the uncut trunks. However, wood seasoned in this manner has a greater tendency to warp than that permitted to dry out more naturally, as had been the practice prior to the twentieth century, it not being uncommon for timber to be left for many months, even years, between felling and sawing. Further, dividing the trunk into planks after the drying process is complete produces planks less prone to distortion, as most of the shrinkage and warping will have already taken place.

The Roman method of processing a tree melds the modern practices with the traditional, drawing the best features from each. With their immediate need for a high volume of usable timber (for construction, weaponry/armour production, fuel and tannery processing), they recognised the value of artificial drying, but minimised the detrimental effects of this by their method of dividing the trunk. In a modern sawmill the trunk is sawn into planks along its length, cutting tangentially to the grain lines. In the Roman period the trunk or 'butt' would be divided into quarters by 'cleaving' – splitting, using either a 'cleaver' or a heavy mallet, or 'beetle', to force iron or wooden wedges along its length. A 'cleaver' is an iron tool with a knife-like blade set in a handle at right-angles to the back of the blade (Hodges, 1976,114), and a 'beetle' (also known as a 'bitel', 'beddle', 'commander' or 'maul') is a barrel-shaped mallet. Its head is often made of apple or elm wood, banded at each end with metal rings to prevent splitting, fitted red-hot and nailed in place with spikes (or 'dags'), and the handle is usually of ash (Sparkes, 1991, 10).

This then allowed air to circulate more freely around a greater surface area during the drying, speeding up the process. After drying, when most of the shrinkage and warping was completed, the quarters were either used with minimal squaring off to remove rounded edges if used for construction, or cut into planks radially, at right angles to the grain lines (J Zant, pers. comm., October 2005), producing planks less prone to distortion than those produced by cutting tangentially (Hodges, 1976, 114).

Where necessary, the 'seasoning' process could then be artificially speeded up by drying or baking the half or quarter trunks in an oven. One innovation introduced by the Romans was the corn-drying kiln. This was not only used for drying grain crops (such as emmer, spelt, barley and oats) as its name suggests, but was also used for seasoning timber (as confirmed by Columella in *De Rustica*, 1.4.19), smoking foodstuffs (fish, cheese and meats), malting grain for beer making, and preparing flax after retting to make rope, cord, nets, fabric, etc. (Morris, 1979, 5–9).

As Vegetius suggests, the Roman military probably aimed to make each fort self-sufficient, dealing as much as possible with its needs within the camp, requiring the storage of large quantities of grain for both men and horses (Vegetius, *Epitoma rei militaris*, 2.10). In order to store grain crops, in bulk or in sacks, it is necessary first to drive off any excess moisture. Grain with around 15 per cent moisture will store for up to eleven or twelve months, but grain with less than 14 per cent moisture could be stored for up to eighteen months. However, grain stored at around 15° C, even with moisture at only 11 per cent, would be prone to mildew, fungal and weevil attack, or could germinate, and this could be prevented by first part-baking to around 49° C in a corn-drying oven (Morris, 1979, 5–9).

A typical corn-drying kiln would consist of a stoking hole, firebox and flue, sunk into the ground, with a drying floor above ground level. It may have been fuelled by charcoal, as in an example from Ribchester (May, 1922, 43), or by a mixture of charcoal and coal, as in the example from Kiveton Park (Radley & Plant, 1969b, 158–161). Other examples of these corn-drying kilns have been found at Templeborough (May, 1922, 43) and Womersley (Buckland, 1986, 35–36).

A later variation, dating from the third and fourth centuries AD, for T-shaped corn-drying kilns can also be seen at Barton Court Farm, Oxfordshire, Huntsham aisled villa building, Herefordshire, and Vineyard Farm, Gloucestershire (Bridgewater, 1965, 182; Rawes, 1991, 84; Dearne & Branigan, 1993, 88). Where these corn-drying kilns are found in association with water tanks, as at Halstock, Hambleton, Huntsham and Whitton, it is suggestive of them also being used in production of malt for beer making, as the barley would need to be steeped in water for several hours, then allowed to germinate before being dried to convert to malt (Morris, 1979, 5–9). However, water tanks are essential in many industrial activities, including the steaming and bending of wood, so this does not preclude any additional industrial use for these ovens.

Sawing

Once the half or quarter trunks were fully seasoned, they would be further divided lengthways into planks. This may have been achieved by cleaving once more with mallet and wedges, or they may have been sawn. In view of the extremely thin planks used in shield manufacture (2–3 mm in some cases), it would be more probable that sawing was used, as this would achieve greater accuracy than splitting.

Traditionally, and even into the twentieth century, planks were produced by sawing over a sawpit. Although there is no positive evidence for sawpits in the Roman period, this does not preclude the possibility of their existence. Pits would be dug up to 5 m long and 2 m deep, either in permanent processing locations or as a temporary arrangement in a clearing in the wood, close to where the timber was originally felled. In this case, a basic framework could have been formed using two logs laid along the top of the pit ('strakes'), with two shorter logs across the end ('sills'), the timber to be sawn being flattened on its upper and lower surfaces with a hatchet to prevent rolling, then fastened to this framework by iron hooks (Sparkes, 1991, 4). However, if the trunks had been seasoned by artificial drying (perhaps in a corn-drying kiln), any sawpit would probably be of a more permanent nature, located close to or inside the fort itself. As an alternative to digging a pit, it is possible that an above-ground framework could have been erected, which would survive archaeologically only as a series of robust postholes.

The planks would be marked out on the trunk or 'butt', using a stretched cord and chalk or charcoal, then sawn along these guidelines by two men using a pit saw. The pit saw has a long blade, tapering to reduce bottom weight, with coarse-set teeth (angled so that the teeth are bent outwards) and a wide kerf (the width of the cut made by the teeth), used in a ripping action, cutting on the downward stroke. It has removable handles at each end, the top handle (or 'tiller') being used to steer the saw, the bottom handle (or 'box') being used by the second man underneath the trunk to push the blade back up. Each plank width would be sawn for a short distance (the end of the trunk bound to prevent vibration and wedges inserted to keep the planks separated and prevent the blade binding), and then the trunk would be pushed further along the frame and the process repeated (Sparkes, 1991, 4).

Other types of saws that may have been used are the 'crosscut', 'bow' and 'frame' saws. The crosscut saw, similar to the pit saw, was operated by two men, one on either side of the trunk, with one removable handle, but unlike the pit saw it was provided with shallow-set teeth to allow it to cut across the grain for cutting into shorter 'butts'. The bow saw in contrast was smaller, intended to be used by only one man. It had an H-shaped frame, with a serrated blade across the base, tensioned across the top, the sides forming the handles. The frame saw, as with the bow saw, was intended for use by one man, the blade tensioned down the centre, within a box-shaped frame, each end forming the handles like a small pit saw (Sparkes, 1991, 6; Hodges, 1976, 116).

Shaping/Finishing
Once the wood had been sawn or split into planks or blocks, it would still require further shaping and finishing. Planks intended for use on shields, for example, would require thinning down to the appropriate overall thickness, thinner towards the edges of the shield, possibly using either an axe or an adze. An adze consists of an iron blade set at 90° to a long handle. The operator would use the adze by standing over the plank, drawing the blade towards him to remove thin slices of wood while holding the plank with his feet, leaving small 'facets' where wood has been scooped away, as described by James on some of the Dura shields (James, 2004, 160).

Further fine smoothing of the surface, or thinning down the thickness of planks used towards the outer edges of the shield, could have been achieved with a plane, or more probably a drawknife, again producing the telltale 'chatter' marks found by James on the Dura shields. The edges would then have been shaved and angled to ensure a snug fit to their neighbouring planks, and the surface smoothed ready for the application of a leather or fabric outer layer, perhaps by using a 'jointer'. The jointer (a tool also used traditionally by coopers) is a blade fixed to the work surface, the wood to be shaped passing over the cutting edge to cut the angle of the bevel (Hodges, 1976, 116).

The 'draw shave' or 'drawknife' is a straight blade with a handle at each end, used by an operator seated on a 'shave horse'. It could also be used to shape rough-hewn blocks of wood, prior to finishing on a lathe, perhaps for shaping shafts for projectile weapons (Sparkes, 1991, 9). The 'pole lathe' consists of a young larch or ash pole, stripped of its bark, shaved on underside for extra flexibility and chained in the upright position to a strong post set in the ground outside the workshop, its end through the eaves of the hut. A cord then passes from the pole, down around a billet of wood on the lathe to be turned, and fastened to the foot treadle. As the treadle is pressed, the wood rotates on mandrels of the lathe, allowing the operator to remove any surplus material, producing a smooth, evenly dimensioned length suitable for balanced projectile weapon shafts (Sparkes, 1991, 9).

LEATHER

EVIDENCE FOR LEATHERWORKING

Leather was the plastic of antiquity, used for a myriad of purposes, moulded to shape, hardened for its protective properties or used in the construction of other objects for its strength and flexibility. It is used extensively in the production of Roman military equipment, for the production of waterproof bags, shield covers, protective undergarments, belts, shoes, thonging, and, less visibly, as the integral strapping holding together segmental ferrous armour. Unfortunately, being a natural product, due to its organic nature it is not often as evident in the archaeological record as its ubiquitous use might suggest. Apart from highly mineralised remains of internal strapping on *lorica segmentata* components (including examples from Corbridge and Kalkriese), and the dubious evidence of leather covering to the Doncaster shield (existing only as a 'bubbly char' which may equally be interpreted as remains of a layer of glue), almost all leather artefacts have originated from waterlogged deposits, mainly outer ditches of forts, rarely as complete items, comprising offcuts of manufacture and discarded fragments resulting from repair work. Of these, it would appear that cattle hide predominates for the use of shoes, and for items requiring large sheets of leather, such as large shield covers (requiring up to two full skins), with goatskin being used for tents (up to seventy skins) and much of the remainder (van Driel-Murray, 1985, 59). In addition, there is the possibility of the use of cattle skin in its untanned state as rawhide, or of boiled leather ('*cuir bouilli*'), perhaps to provide hard, resilient edges for shields (as this retains its shape and can be moulded, then coated with wax, resin or pitch to waterproof), although there is no archaeological evidence for its use in armour (Hodges, 1976, 152).

Skins and hides would have been used from the earliest periods, as a natural by-product of hunting meat for food. After death of the animal, decay and bacterial action would soon lead to putrefaction of the hide unless some level of processing was carried out. Hides and skins were probably first used almost untreated, straight from the slaughter/butchery processing, just scraped off excess meat and dried in the air and sun. These would have rotted quickly in the heat and stiffened when cold or wet (effects partly minimised by rubbing with animal fats – methods documented by both the Assyrians and also by Homer in his *Iliad*). Homer compares the desecration of the dead on the battlefield to the stretching and oiling of a great ox hide (*Iliad*, 17.389–95):

> As when a man gives the hide of a great ox to the folk for stretching, once he's soaked it with oil, and they pick it up and, leaning apart, they stretch it in a circle; the moisture springs out, the oil soaks in, with many tugs it's all stretched out between – so this way and that they tugged in a little space at the corpse on both sides.

The hides would then have been soaked to re-soften them, probably drying over a fire (accidentally causing them to absorb resinous materials and aldehydes from the smoke and wood ash), which may have led to the accidental discovery that twigs, bark and leaves soaked in the water would have helped to preserve them (Mastrotto, 2007; Fernando, 2007; Hodges, 1976, 149).

Evidence for the early use of skins and hides being processed as leather (artefactual and traditional) can be seen throughout Europe, for example in the clothing worn by the Alpine huntsman, dating to 3300 BC, held in the South Tyrol Museum of Archaeology (Demetze, 1998). (The body of a Chalcolithic huntsman, known as 'Oetzi', or 'Ötzi' after his find place, was discovered in September 1999, on the Schnalstal glacier, in the Ötzal Alps, between Austria and Italy. Several of his personal belongings were also found with him, along with items of clothing, many being crafted in leather, and stitched with animal sinews, including a hide coat, cap, and shoes made of deerskin, with bearskin soles.) Evidence can also be found in the Mongolian tradition for leather flasks, blankets, clothing and headwear, oil-tanned to waterproof them.

Further evidence (artefactual, sculptural and written) can also be seen in the Near East (Assyria, Mesopotamia/Sumeria and Egypt; Mastrotto, 2007; Fernando, 2007). Sculptural depictions (and some later archaeological examples) show Sumerian people wearing long leather dresses and diadems from the fifth millennium BC. Similarly, the Assyrians are known to have used leather for shoes and wineskins (which they used inflated as floats to support timbered craft, in addition to leather-coated, wicker-framed coracles, used on the rivers above Babylon; Herodotus, *Histories*, 1.194). The Egyptians also used leather for clothes, tools, weapons, etc., with sculptural evidence from stone carvings of leather workers from the fourth millennium BC and artefactual evidence in the form of sandals from 1300 BC and a queen's funeral tent from 1100 BC (made of gazelle skin). The Israelites are also known to have used leather, a tradition possibly predating – or possibly learned from – Egypt, which is documented in the Bible ('Unto Adam and also unto his wife did the Lord God make clothes of skin'; Genesis 3:21), and the first-century AD geographer Strabo tells of the Phoenician's use of leather water pipes (Strabo, *Geog.*, 16.1.6).

In the Greek and Roman Mediterranean, there is mention of the use and preparation of leather from both Homer and Herodotus (*Histories*, 1.88), in addition to numerous artefactual remains and the physical evidence of a first-century AD tannery from Pompeii. There are numerous references in Homer's *Iliad* to the use of leather (in addition to that for the preparation of hides from Book 17 previously quoted):

- Book 7 describes how Tychius, 'by far the best worker in leather', made Ajax's shield of bronze, covering an underside made from seven layers of hides from 'full-fed bulls'.
- Book 10 describes Odysseus wearing a 'helm of leather'.
- Book 12 describes how the Danaans repaired their parapets with leather hides; Lycians using leather bucklers; and Sarpedon carrying a bronze shield, its 'inside formed of leather'.
- Book 14 describes a swineherd cutting out sandals from 'good, stout ox hide'.

Evidence of methods employed in such a tannery comes not only from the physical remains of the fixed vats in the workshop itself, but from writings of Pliny the Elder (who lived contemporary to the time of operation, his demise being due to the same eruption which preserved the workshop remains), who describes a recipe for a tanning solution containing 'bark, gallnuts ['gallnuts' are hard, round growths, high in tannic acid, formed as a defensive measure by oak trees, in answer to attack from insect eggs laid on their leaves and buds], sumac and lotus' (Pliny, *Nat. Hist.*, 13.19; 23.6). Further linguistic evidence can also be drawn, in that the word 'pecuniary', meaning 'money related', derives from the Latin term *'pecus'*, meaning 'cattle', suggesting the use of cattle and/or hides as an early form of barter, or currency (Fernando, 2007).

PROCESSES

Hide/Skin – The Raw Materials

Animal skin is made up of three main layers: the epidermis, dermis and hypodermis (subcutaneous fatty tissue). The hypodermis is the lowest layer, closest to the flesh of the animal,

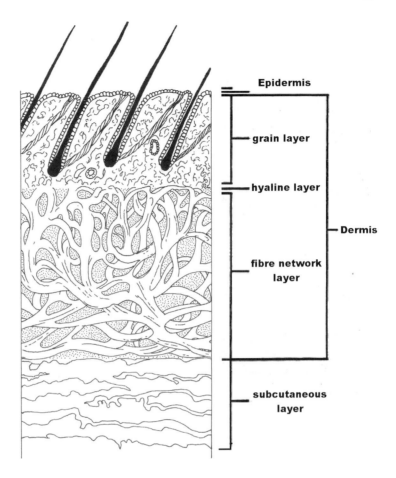

Epidermis

grain layer

hyaline layer

Dermis

fibre network
layer

subcutaneous
layer

Fig 78. Different layers
of tissue in a skin or
hide. (Artwork by J. R.
Travis, from Koninklijke
Bibliotheek, 2007)

consisting of a band of subcutaneous connective tissue, globules of fat, nerves and larger
blood vessels, which will quickly putrefy if not removed once the hide or skin is removed
from the carcass. Similarly, the epidermis, the thin outer layer of scaly cells, also needs to be
removed prior to preservation and use (Fig. 78).

Between the epidermis and the hypodermis, the dermis is the thick inner skin of fat cells,
fibrous water proteins (collagen), carbohydrates (glycogen) and minerals, forming sweat
glands, hair follicles, sebaceous glands, nerves and small blood vessels (Hodges, 1976,
148–152). Collagen consists of long-sequence amino acids (polypeptides) which form into
long, fibrous parallel chains, giving it elasticity. Again, the dermis can be seen to consist of
three layers: the grain, the junction and the corium. The grain is the top layer, closest to the
epidermis, where the collagen fibres are smaller and more densely packed and the follicles
are most visible. The corium is the lowest part, forming the bulk of the dermis layer, where
the collagen fibres form large, strong bundles. The junction layer, between the grain and the
corium, is weaker than both as it may contain globules of fat, particularly evident in sheep
hides, which, when dissolved during the tanning processes, can leave voids that cause peeling
and create a spongy texture (Kelly, 2004, 2–3).

Processing Hides and Skins
Hides are removed from the animal by 'flaying' soon after slaughter, 'fleshed' to remove
residual meat and fat, then either processed immediately, or 'cured' by covering in salt grains,
or soaking for at least sixteen hours in large pools ('raceaways') of salty water (brine) to
prevent any deterioration or putrefaction. Treated in this way, 'cured' hides could be stored
for several months in cold rooms, or transported to tannery sites for more permanent

treatment into leather (Fernando, 2007; COTANCE, 2007). It is not archaeologically apparent whether one of these methods was preferred exclusively during the Roman period, although van Driel-Murray reports shallow depressions visible on leather offcuts found at Xanten and Velsen, which she interprets to have been caused by salt having been scattered on hides as preliminary processing at slaughter sites prior to subsequent tanning, perhaps elsewhere (van Driel-Murray, 1985, 61).

Soaking

Before use, the cured hides needed to be rehydrated, by soaking in a vat for several days to remove water-based minerals and salt, washing off blood and dirt. This time can now be shortened to around ten to twelve hours in modern processing sites by agitating in a rotating drum, although there is no evidence for this method being employed during the Roman period (COTANCE, 2007; Kelly, 2004, 4; Fernando, 2007).

Dehairing – 'Sweating' or 'Liming'

After soaking, the hide needs to be stripped of its hair and the flaky epidermis layer in a process known as 'sweating'. Although in a modern tannery this process is executed using synthetic chemical preparations, natural products were previously utilised. However, in either case, the hide is soaked in a strong alkaline solution to induce a controlled chemical decay, partially dissolving the epidermis and the keratin in the hair, and opening up and 'plumping' without damaging the collagen fibres. This strong alkaline would have been produced from fermented urine, or fresh wood ash, with the introduction from the middle ages of rubbing the moistened surface with quicklime, lending the process the alternative name of 'liming'. The decayed hair and epidermis can then be removed with a 'curriers comb' (a series of teeth set in a circle) or scraped away with a blunt knife (Hodges, 1976, 149; Kelly, 2004, 5; Mastrotto, 2007).

Defleshing/Scudding/Deliming

With most of the epidermis layer and surface hair removed, the hide or skin is then soaked in water, laid over a bench and pounded to remove any residual hair. It is then scraped to clean the underside of the skin, smooth irregularities in thickness and remove unwanted subcutaneous fatty tissue. Initially in early periods this would have been possible with stone tools, but later leatherworkers would have used a drawknife (also known as a scudding, fleshing, beaming or paring knife; Hodges, 1976, 148; Kelly, 2004, 5). Once the hair has been removed, prior to dressing/tanning, the hides or skins would be treated with a weak acid, enzymes and salt to neutralise alkalinity from the sweating/liming process, but without dissolving the natural buffer of calcium salts (COTANCE, 2007; Fernando, 2007; Kelly, 2004, 5).

Dressing

Once the fat and hair has been removed from the hide (or 'pelt' as it may now be known), it needs to be preserved to prevent stiffening and bacterial decay, or putrefaction. This can be achieved by either reversible methods ('dressing') or by irreversible, permanent methods ('tanning'), although the former may be performed as a precursor to the latter, as leather which has been 'dressed' alone would be prone to decay, its preservative effects being removed by subsequent soaking (deliberate or accidental).

There are three methods of 'dressing' that are known to have been used: smoking, oil dressing, and salt/alum dressing or tawing. In 'smoking', the hide is suspended over a slow-burning wood fire or in a special chamber, the hide absorbing the resinous materials and mixture of aldehydes produced by the wood during burning. In 'oil dressing' or 'chamoising', the hide is rubbed with a mixture of oils (such as tallow, egg yolk, cattle brains, etc.), all of which being fats that oxidise in the air. The third method, 'salt dressing', 'alum dressing' or 'tawing', uses dry mineral preservation, rubbing salt and alum (a mineral derived from volcanic areas) into the hide, producing a leather which is very pale in colour (known by the

Romans as '*aluta*'), but the low water resistance of which made it unsuitable for clothing, tentage, etc. As an alternative to alum, it would also have been possible to use wood ash, but with inferior results (Hodges, 1976, 149–150; Kelly, 2004, 6).

Tanning
Tanning chemically and structurally modifies hide and skin permanently, irreversibly turning it into leather (in that it can be boiled for up to three minutes without shrinkage; Hodges, 1976, 150; Fernando, 2007). This can be achieved by two methods: either vegetable tanning or, most commonly from the mid-nineteenth century to the modern day, chemical tanning (for example chrome tanning). In preparation for tanning, the hide must first be 'pickled' or 'bated', by soaking it in an infusion of warm water, dung, dog excreta, etc. This process, also known as 'plumping' or 'puering', removes some of the muscle fibres and softens the hide, making it swell through partial putrefaction (bacterial action) until 'flabby' in appearance (Hodges, 1976, 150; Fernando, 2007; Kelly, 2004, 6). In more modern tanneries, of course, the dung would now be replaced by salt and hydrochloric or sulphuric acid (Fernando, 2007).

Chrome or Chemical Tanning
In chrome or chemical tanning, the hides or skins are rotated in a drum, bathed in water, dyes and synthetic tanning agent (trivalent chrome), to temperatures of 49° to 60° C, for up to eight hours, until the chrome soaks into the skin. This is then fixed by the addition to the solution of alkalines, such as sodium carbonate or bicarbonate, and finally the skin/hide is lubricated with natural or synthetic fats, before finishing with waxes, synthetic plastics, etc. (Fernando, 2007).

Vegetable Tanning
In vegetable tanning, after bating/pickling, the hide is then moved to a tanning bath, moving through successively stronger mixtures over time, to allow the solution to penetrate the hide completely. This is a much slower process than the modern 'chrome' tanning method (taking two to four days if the mixture is agitated artificially, for example in a rotating drum, or several weeks/months by more conventional methods), and involves the use of natural products, 'tannin', or tannic acid, generally being derived from plant material such as oak galls, the bark of trees and leaves (Hodges, 1976, 150; Fernando, 2007; Kelly, 2004, 6).

There are two types of tanning mixtures, each giving different colour leathers:

• Pyrogallol or hydrolysable. These mixtures are formed by esterification, which is an infusion of the woods and bark of oak, chestnut, acorns, etc. This method would have been the most common used in the Roman period, producing pale-coloured leathers.
• Catechol or condensed. These are mixtures are formed by condensation, using a variety of woods, including mimosa, larch bark, etc. This method produces dark-coloured leather, although it was probably not widely used in the Roman period.

Collagen is a water protein. That is to say that water exists within its molecules and also between the fibre bundles. When skin dries, some of this water is lost, permitting the fibrous polypeptide chains to hydrogen bond together, reducing flexibility, a process which is reversed when rewetted. During the tanning process, tannin replaces the hydrogen-bonded water, inhibiting cross-linking of the collagen chains, stabilising the dermis, making it water resistant, supple and more durable (Kelly, 2004, 7).

As with 'bating' or 'pickling', this process would usually require the provision of vats or tanks, which may be noticeable in the archaeological record. However, it is possible to produce a similar effect in small-scale production, albeit less efficiently, by 'bag tanning', whereby the skin to be processed is sewn up, suspended and then filled with the tanning solution, which then slowly permeates through over time (van Driel-Murray, 1985, 62).

Samming, Splitting, Skiving and Neutralising
After tanning, the hide undergoes 'samming', where excess water is removed from the hide or skin. The leather is then reduced to an even thickness and thick leather may be split to produce several sheets of thinner leather, the top layer being the 'grain' leather (which can be brought to an even thickness and any irregularities removed from the underside during a process known as 'skiving'). The lower layers can then be processed as suede, or in the case of calfskin, for parchment, or vellum (which is produced from split thin layers of calfskin, oil dressed to make it translucent). The sheets of leather would at this point still be slightly acidic, from having been soaked in the baths of the tanning solution, and would need to be washed to neutralise any residual acid (Hodges, 1976, 151; COTANCE, 2007).

Drying
After tanning, the leather needs to be dried. In modern processes this would be done by vacuum, but in more primitive societies it would have been carried out by pegging down the hide flat to the ground (COTANCE, 2007). However, this latter method is not very efficient, as moisture trapped on the underside of hide can provide conditions for bacterial and fungal attack, spoiling the hide. An improved method, used in more 'modern' societies, and a method suggested to have been used in Roman tanneries, involves stretching the hide between pairs of upright posts, providing better all-round ventilation, with quicker and more even drying (van Driel-Murray, 1985, 63). Similarly, gentle drying may have been possible by hanging in ovens, such as the multi-purpose corn-drying kiln (used for everything from corn drying, as the name suggests, to curing wood, leather-drying pottery prior to firing and drying ores prior to smelting).

Vegetable-tanned leathers produced in this manner would be strong enough to support the component parts of segmented plate armour, and sufficiently waterproofed to serve for tentage (panels of which are evident from exterior ditch deposits at many Roman forts, including Ribchester, Vindonissa and Vindolanda) and removable covers for shields (as found at Vindonissa and Castleford; Figs 61–63; van Driel-Murray, 1985, 53; 1989, 18–19; 1999, 45–54) and as described in the writings of both Caesar and Dio (Caesar, *Bell. Gall.*, 2.21.5; Cassius Dio, 61.3). As the tanning process would have made the leather more or less impermeable to water, it is unlikely that these covers would have been coloured, as the pigments or paints would not have adhered (in modern leather processing and 'finishing', the hide has to be mechanically 'staked' to make the surface permeable, in preparation before colouring, embossing and ironing to the required texture, shiny or matt; COTANCE, 2007), surface designs of leaves, petals and legionary symbols being added by way of appliqué patches, presumably of contrasting shaded leather (van Driel-Murray, 1989, 19). In contrast, however, the 'leather' or 'skin' coating applied to the surfaces of some of the Dura shields does appear to have been painted, and it is possible that for this purpose, alum-tawed leather may have sufficed, its less impervious structure permitting the paint to better soak in, with the layers of gesso and paint above supplying the necessary waterproofing (James, 2004, 163).

GRAIN

Leathers of different animals have different grains, from the arrangement of hair follicles and sweat glands, and the coarseness and the arrangement of their collagen fibres, with the finer leathers deriving from smaller animals (for example goats, sheep, small deer), or the young of large animals (as with calfskin; Hodges, 1976, 151). Hides are produced from larger and adult animals, such as cow, horse, pig, sheep and goat, and skins are produced from small and young animals, for example kid, calf or lamb (Kelly, 2004, 3; Fernando, 2007). The arrangement of the hair follicles on the grain of leather artefacts can therefore be helpful in identification of materials used in their construction. Calfskin is smooth and even textured with small, regularly spaced follicles, whereas pigskin is stiffer with larger, widely spaced follicles in groups of three. Sheepskin and goatskin are more difficult to differentiate, however, as both animals are closely related, although goatskin in comparison is firmer, with a more compact grain layer. In contrast, sheepskin is weaker and less durable, with a spongy texture caused by the voids created by wool fat (lanolin) within the dermis, and its looser collagen fibres. However, within the range of sheep there are two distinct types – hairy and woolly sheep, the skins of the hairy sheep being of better quality than those of the woolly sheep, possibly to the extent of being visually almost indistinguishable from goatskin (Kelly, 2004, 3).

These similarities between goat and sheepskins, and the potential for misidentification, is discussed by van Driel-Murray in consideration of the Roman military supply network for the preparation and movement of hides, suggesting that hides may have been shipped with horns and hooves still attached, to reassure the leatherworker that an inferior product (sheep) had not been substituted in his order for goat (van Driel-Murray, 1985, 60).

ARCHAEOLOGICAL EVIDENCE FOR ROMAN LEATHER WORKING

Leatherworking is an activity that would not be reflected in the solid archaeology of buildings, as it does not require any special features, in contrast to buildings used as workshops for metalworking (such as hearths, water tanks, lathe pits, etc.), for woodworking (saw pits, lathe pits) or tanneries (water tanks, soaking vats, drying rooms, etc). Leatherworking can therefore be carried out almost anywhere, providing lighting is adequate and supply of raw material can be provided (and with the Roman supply infrastructure, that is virtually limitless). The only diagnostic evidence for leatherworking would come from tools (and a good workman would not discard these lightly), or from the detritus of the activity (such as offcuts of 'new' leather from manufacture of new goods, or discarded panels and fragments cut from older items during regular maintenance (Figs 79–82). Crescent or half-moon-shaped blades would have been used to cut the leather, being superior to pointed knives, which soon blunt. Sewing would have been carried out with metal or bone needles, using thong, animal sinew or thread, the stitching holes being made by an 'awl' (Hodges, 1976, 152). From the shape of stitching holes in discarded tent panels from Ribchester, the awl used would appear to have been of flat or diamond-shaped section.

Although the majority of Roman sites with any waterlogged deposits will produce finds of discarded fragments of leather objects, such as tents, shoes, belts, etc. (as part of the detritus of everyday life), only finds of offcut fragments of 'new' leather will be indicative of leatherworking craftsmen. Carol van Driel-Murray has made a study of a number of Roman sites with evidence of these offcuts, subdividing into three categories finds that may suggest the nature of the site and type of activities being carried out (van Driel-Murray, 1985, 49).

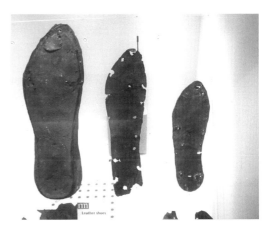

Fig 79. Leather panel from Ribchester, possibly from a garment or more probably from a tent. (Photograph by J. R. Travis)

Fig 80. A selection of shoe soles from Ribchester. (Photograph by J. R. Travis)

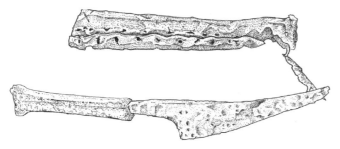

Fig 81. Knife from Ribchester, possibly for leather working. (Artwork by J. R. Travis)

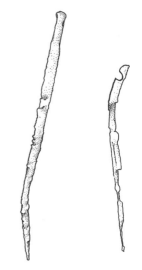

Fig 82. Leatherworking needles from Ribchester. (Artwork by J. R. Travis)

The first category of fragment is the primary offcuts: the unusable and poor-quality parts of the hide which were probably trimmed away from the hide at the primary processing site – probably the tannery itself. The presence of unusable waste parts, for example at Xanten, Maastricht and Vindonissa, suggests hides arriving at the tannery (possibly located at a processing 'base camp') with head, tail and udders still attached, with only minimal processing at original slaughter sites. The trimmed and processed hides would then be taken to workshops (at local or remote sites) for secondary processing into end products, the removed excess minimising bulk to be transported.

The second category of fragments described by van Driel-Murray is those secondary offcuts produced when a pattern is cut from the hide, producing the rough outline of a defined article, for example offcuts of single-piece shoes from Maastricht, and soles from Velsen, indicating a shoemaker's workshop rather than a tannery site. From piecing together offcuts found, it can also be seen how economically the hides would be used, carefully fitting pattern pieces around each other to minimise wastage, although some of the scraps so produced are at least large enough to show the grain of the leather to indicate animal type. From similarities in shoe design (and even nail patterns) seen across a number of first-century AD sites, the suggestion is for standardisation of military issue at that time.

The third category of fragments is the small trimmings from the basic pattern pieces. Where these are found they are indicative of final assembly and finishing of the end product, although not necessarily the work of the skilled craftsman. Van Driel-Murray proposed that the cutting of the basic pattern may have been the work of a skilled craftsman, following a standard pattern, who then supervised the assembly by unskilled workers, citing by example a work roster from Vindolanda suggesting that a third of the fort's workforce was involved in shoe production (van Driel-Murray, 1985, 54).

She proposed, then, that hides were produced as a by-product of meat, with forts being mostly self-sufficient to their own needs. The roughly prepared hides would then be subjected to primary processing, by salting immediately after slaughter to preserve them, suggested by salt marks seen on offcuts from Velsen and Xanten (van Driel-Murray, 1985, 61). The hides would then be transported back to the nearest base camp for trimming and tanning, before being returned (or sent to other sites where need was greatest) for manufacture into leather goods. These 'goods' may have been just shoes (a basic requirement at every fort, whether auxiliary or legionary), or could be something a little more specialist (shield covers, tents, etc.), which only appear to have been produced at sites capable of larger-scale production (for example, Vindonissa), although all sites would have carried out everyday maintenance and periodical repair on such items. For example, at the site at Valkenburg, offcuts of 'new' leather and scraps of repaired items suggest the location of a shoemaker's shop in a wing of the *principia* involved in small-scale manufacture and repair, with two further buildings associated with leather articles interpreted as 'stores', and with specialist goods and raw materials supplied from a base camp, such as Vindonissa (van Driel-Murray, 1985, 54; 67, Fig. 8).

The apparent uniformity of design in tents, shield covers and military boots, which could be seen in the first century AD in 'periods of forward movement', particularly along the forts of the limes, begins to become less distinct in later periods, with the disappearance of the three-piece military boot in favour of closed, 'civilian'-type designs in the early second century AD, accompanied by a fall in quality, and with no shoe factories being evident by the fourth century (the time of the *Notitia Dignitatum* lists of known *fabricae*). Van Driel-Murray suggested that this trend may be due to shoemaking at least moving from military control into the civilian private sector, with this privatisation of shoe supply in 'periods of consolidation' being a logical trend, deriving from the a need to meet the high capacity of shoes required by the military, estimating that all military personnel would have required replacement footwear at least every six months (van Driel-Murray, 1985, 58; 68, Fig. 9).

In the sites of secondary processing (leatherworking as opposed to tannery), leather offcuts often describe the shape of the object being produced, appearing as a negative of the pattern shape cut from the hide, and this may provide clues to the nature of goods being produced, for example the fragments from Maastricht indicate the manufacture of single-piece, civilian-style footwear, suggesting either supply intended for a civilian market, or civilian-style military supply, possibly being produced by a civilian workforce (van Driel-Murray, 1985, 48). In addition to leather offcuts, evidence of the manufacture of leather goods may also come from the presence of goat horns and remains of hooves (metapodials), as in the case of goatskins these would have been left on the skin before transportation to secondary (manufacturing) sites (to avoid confusion with physically similar but inferior sheepskins, as previously discussed), in contrast to the processing of cattle hides, where horns and hooves would be removed at the primary slaughter or tannery sites to minimise damage during shipping (van Driel-Murray, 1985, 62).

ARCHAEOLOGICAL EVIDENCE FOR ROMAN TANNERIES

Tanning is not necessarily always carried out at the same location as the slaughter of animals for meat, the flayed skin probably undergoing some immediate minimal cleaning and processing to avoid putrefaction, prior to transportation to a more specialist processing site, where it could be stabilised and permanently converted into the more durable and water-proof product, leather.

In addition to the presence of primary offcuts of unusable waste material (such as the heads, tails, udders of cattle and the poorer-quality leather from the edges of hides), suggested as evidence of tannery activity by van Driel-Murray (in contrast to secondary leatherworking), the tannery process may also leave some more tangible archaeological features within buildings, such as water tanks and fixed vats for soaking hides, as were found both in the tannery complex at Pompeii and also in the fort at Hofheim (van Driel-Murray, 1985, 62).

However, the Pompeii site is unusual for two major reasons. Not only does the town of Pompeii afford a unique insight into all areas of Roman life at a distinct moment in time, including these less glamorous aspects of industrial activities, but also it is doubly unusual that these activities are found at all in such close proximity to general habitation. In view of the more unpleasant materials used in the tanning process, most tanneries would, quite reasonably, be expected to be located at some distance from a settlement or military instal-lation, preferably downwind. Consequently most tannery sites are not usually pinpointed, their existence only being implied by deposits of waste products from their activities, for example, the primary offcuts, as described being found at sites such as Xanten, Maastricht and Vindonissa (van Driel-Murray, 1985, 49), or finds of cattle hooves, horns and horn cores, as described from Hofheim or Velsen (van Driel-Murray, 1985, 64). It is also possible that some small-scale tanneries may have operated without need of large, permanent vats, using the alternative method of 'bag tanning', whereby the sewn-up hide is filled with the tanning liquor and suspended, allowing the solution to permeate the skin from the inside. The only evidence then of this activity may possibly be in the form of holes for these suspension posts. Similarly, paired post holes may be indicative of where hides had been stretched and dried, a more efficient method of drying than the primitive pegging down (van Driel-Murray, 1985, 63).

APPENDIX V

METAL

EVIDENCE FOR METALWORKING

Metalworking was no new innovation of the Roman period, with cold-worked nuggets of naturally occurring soft metals such as copper and gold dating back to the early Bronze Age. Later, in addition to the exploitation of the more precious but less practical metals, the hardening properties of natural impurities in copper can be seen being manipulated in a deliberate alloying of metals, possibly as a result of observation over time of the use of sulphide ores high in arsenic content, with hardened copper objects being produced from before the third millennium BC, and tin bronzes appearing in central Europe from around 1800 BC (Champion, *et al.*, 1984, 198–213). Initially, bronze objects were small and mostly produced by use of simple casting methods, although as casting techniques became more sophisticated, allowing secondary casting on to preformed objects, with the introduction of the improved tin bronze, sheet bronzes began to be used for larger objects, such as food vessels, shields and armour (Champion, *et al.*, 1984, 215–284).

The bronze produced was not exclusively used close to the ore's point of origin, with evidence of trade in bar ingots from the second millennium and recycling of scrap metal to established industrial sites, evidenced by numerous shipwrecks carrying cargoes of scrap along the coastal routes from Spain to France and Britain, who appear to be foremost during the later Bronze Age for both quantity and quality of bronze working, being credited with the innovation of the development of copper-tin-lead alloys from the tenth century BC. For example, finds of two bronze Etruscan Negau-style helmets (dating to the fifth and sixth centuries BC) were found from the Les Sorres shipwreck site (dating to the second century BC). Although in remarkably good condition, these were not of contemporary styles and did exhibit evidence of wear, suggesting that they must have been of considerable age at the time of the wreck, their presence being interpreted as part of a cargo destined for recycling (Izquierdo & Solias Aris, 2000, 1–11). The addition of 7–10 per cent lead to alloys of copper and tin produces a slightly softer compound, but one that pours more easily, with less casting failures, therefore increasing productivity.

From the tenth century BC, a new metal, iron, began to appear. This was more difficult to produce, requiring a higher level of technology, but with its greater strength and hardness it began to replace the use of bronze for tools, weapons and armour (Champion, *et al.*, 1984, 226). The Romans then built on these long-standing traditions, with the introduction of new, innovative techniques to increase production and improve the quality of the metal, particularly, and not entirely coincidentally, at times of expansion and radical change within the structure of its military, in answer to a need for large quantities of mass-produced armour at short notice.

Although some use was made of the less practical decorative metals such as gold and silver, for highest-status ceremonial officer equipment (whose armour was less likely to ever see serious heavy combat), the majority of military equipment was made from iron or copper alloy, and for this reason further discussion shall focus mainly on these two metals.

Mining and preliminary processing for both iron and copper were not completely dissimilar, although, as stated, iron requires higher temperatures and more technologically sophisticated furnaces. Primary processing of ores was probably carried out close to where it had been extracted (as suggested in contracts, such as '*Lex Metallum Vispascense*', one of two bronze tablets found on slag heaps at Vipasca, Aljustrel in Portugal; Travis, 2005), and then transported to secondary processing sites (probably in industrial complexes) using road, river and coastal systems. Apart from substantial deposits of iron ore in Britain (possibly a major factor influencing the Roman expansion into this country), the Romans also mined iron in many other provinces (such as Gaul, Spain, Italy, Elba, Sardinia, Sicily, Central Europe, Noricum, Illyria, Macedonia, Asia Minor and Africa; Apollonius of Rhodes, *Argonautica* 2.1001–1007; Healy 1978, 64–65; Davies, 1935, 67, 76, 86, 152–153, 165–173). Ore was extracted mostly using 'open cast' methods ('bell pitting' of deposits close to the surface, and quarrying of outcrops from between agricultural field terracing), and the collection of nodules and pebbles from riverbeds, but with the possibility of some deeper mining from 'adits', as seen at Lydney Park (Tylecote, 1986, 155).

PROCESSES

Iron Working
Iron ore exists in four main forms:

Siderite, chalybite or spathic ore ($FeCO_3$) – iron carbonate, containing 48.3 per cent iron. This is the most commonly occurring form of iron ore, found in nodules of clay ironstone below coal seams and surface outcrops (such as the Wealden Series), or in some sedimentary deposits (as in Northamptonshire, Lincolnshire, Oxfordshire and Cleveland). However, it requires more roasting prior to use than other types to remove excess carbon (as carbon dioxide – CO_2).

Limonite ($FeOOH$ or $Fe_2O_3H_2O$) – iron oxyhydroxide, containing 60 per cent iron. This is a hydrated form of iron oxide. It is less common than the carbonate ores, but higher in iron and requiring less roasting processing (see below). It can be found in the Forest of Dean; Weston-under-Penyard (a major production centre at *Ariconium*, with slag heaps covering a 200-acre site; Jones & Mattingly, 1990, 193), and Lydney Park (evidence of deep mining from an underground 'adit' mine, sealed beneath a hut dating from the late third century; Wheeler, 1939; Davies, 1935, 153; Tylecote, 1986, 155); and on the South Wales coalfields.

Haematite (Fe_2O_3) – iron oxide, contains 70 per cent iron. Again, this is less common than the carbonate ores, principally occurring in the west and south of the Lake District of England (West Cumberland) and Furness (Northwest Lancashire), but the ore highest in iron content and requiring less roasting processing (see below).

Iron Pyrites or Marcasite (FeS_2) – sulphide ore. Common ore found in sedimentary rocks, and also another less frequently occurring sulphide ore, **Pyrrhotite** (Fe_7S_8). Not widely exploited due to presence of sulphur, which can cause brittleness in iron produced from this ore. Often found in varying amounts as an impurity in coal, which can again lead to embrittlement of the metal during the smelting process if high-sulphur-content coal is used (Tylecote, 1986, 225).

During iron production the ore passes through various processes. Firstly the extracted ore is crushed and washed, then roasted to remove carbon dioxide, moisture and excess lime. Next, the roasted ore is smelted in a furnace, in a reducing atmosphere, to produce a 'bloom' of metal. This bloom is then reheated to 900° C and hammered to remove impurities, trapped slag and charcoal (which would otherwise cause it to be brittle), and formed into soft, low-carbon wrought iron 'pigs' (approx. 10 cm and 7 kg), during primary 'smithing'. It may then be transported considerable distances prior to secondary processing from the 'pig' into useable items ('forging').

Roasting
Before smelting, the ore (particularly carbonate ore) is roasted to remove carbon dioxide, excess water, lime, etc. This increases reduction to metallic iron, weakens nodules and helps very hard ores to fracture. Combined with washing, this removes some of the unwanted impurities within the ore, known as 'gangue'.

Carbonate ores: two stages of chemical change
$$FeCO_3 \rightarrow FeO + CO_2$$
Thermal decomposition

$$4FeO + O_2 \rightarrow 2Fe_2O_3$$
Oxidation

Limonite ore:
$$FeOOH \rightarrow Fe_2O_3 + H_2O$$
Removal of excess moisture

An indication of roasting taking place is that the roasted ore changes to a magenta colour. High temperatures are not necessary for this process ($300°$ C to $400°$ C is sufficient), and it may be possible to use an open fire, although shallow pits are better, the temperatures being more efficiently achieved through use of bellows (Bestwick & Cleland, 1974, 143–144). A variety of fuels are possible at this stage without detrimental effect to the structure of the metal, Cleere suggesting 1:1 ore:charcoal, although Tylecote considered that charcoal was unnecessary, and that it would be possible to use coal or wood more economically (Tylecote, 1986, 156).

Three types of roasting hearth are known: pit, stone-lined trough, or lime-kiln type (with possible evidence of some reused limekilns). Some pit-type hearths have been found at Bedford Purlieus and Great Casterton (Fig. 83; Tylecote, 1986, 155, Fig. 94), which are similar to smelting furnaces but much larger, being up to 2 m in diameter. Stone-lined trough, or trench-type roasters, as found at Bardon in the Weald (Fig. 85; Cleere, 1970; 1971, 203–217; Tylecote, 1986, 155, Fig. 95), formed the basis of those used by Cleere in his experiments at Horam (Tylecote, 1986, 156). Examples of keyhole-shaped, limekiln-type roasters can be seen at Cirencester (Tylecote, 1986, 156, Fig. 96), Wilderspool (Fig. 84; Tylecote, 1986, 156–157, Fig. 98) and Corbridge (Tylecote, 1986, 156–157, Fig. 97; Fig. 86), some of which may have previously served for the production of lime for building mortar and other processes. More evidence of roasting hearths has also been found at other sites, such as Bradfield, Conisborough Park (unpublished; Travis, 2005) and more recently at Vindolanda (R. Birley, pers. comm., August, 2005).

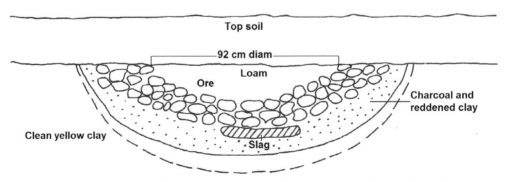

Fig 83. 'Pit'-type roasting hearth from Great Casterton. (Artwork by J. R. Travis, after Tylecote, 1986, fig. 94)

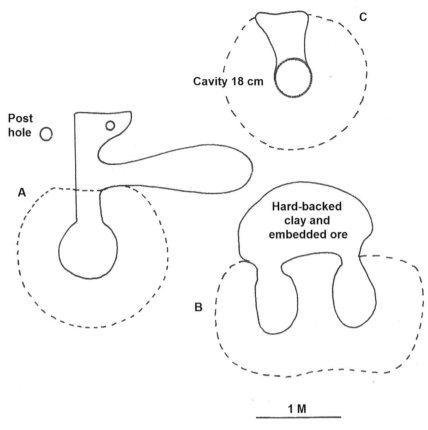

Fig 84. Furnaces from Wilderspool probably used as roasters. (Artwork by J. R. Travis, after Tylecote, 1986, fig. 98)

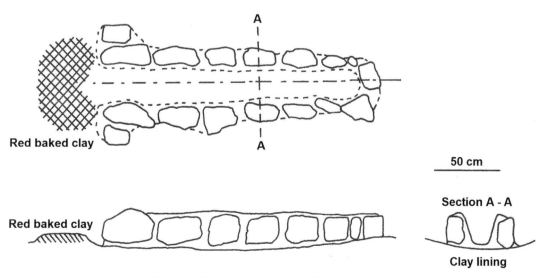

Fig 85. Trench-type roaster from Bardon, Sussex. (Artwork by J. R. Travis, after Tylecote, 1986, fig. 95)

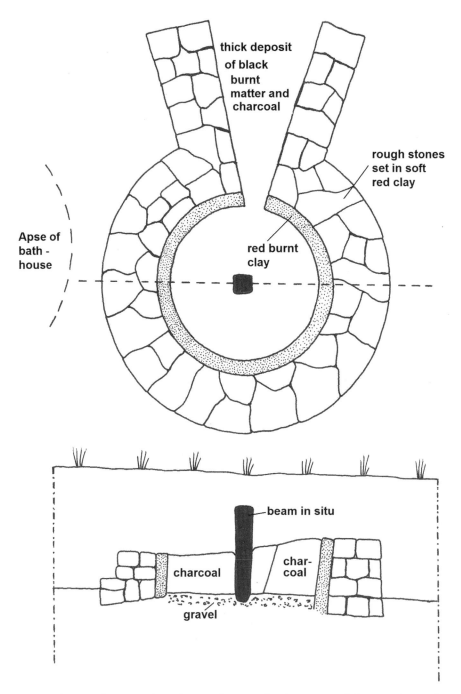

Fig 86. Limekiln reused as a smithing furnace from Corbridge. (Artwork by J. R. Travis, from Tylecote, 1986, fig. 97)

Smelting

Smelting involves heating a charge of roasted ore and fuel in a reducing atmosphere to temperatures between 800° C and 1300° C. Carbon monoxide is produced from a combination of oxygen (from the air draught) and fuel (charcoal or coal), its presence indicated by a change to blue colouration in the flame. At temperatures in excess of 800° C, this then reacts with the roasted ore (Fe_2O_3) forming iron bloom (Fe) and carbon dioxide (CO_2).

$$Fe_2O_3 + 3CO \rightarrow 2Fe + 3CO_2$$

Impurities within the ore (mainly silica) combine with other materials acting as fluxes (such as lime, dolomite or magnesia), often added deliberately to encourage this process, producing a glassy, non-metallic waste product known as slag (often the only diagnostic evidence found archaeologically of the smelting process), which consists mainly of the compound fayalite ($2FeOSiO_2$). However, temperatures of around 1150° C are required for this waste product to liquefy and separate from the metallic iron bloom. At this point most runs off as 'tapslag', with some remaining inside the furnace as 'cinder', and a small quantity still trapped within the 'bloom' which will need to be removed by hammering during secondary processing (Bestwick & Cleland, 1974, 143–144; Table 12).

Coghlan (1977) suggested three types of smelting furnace were used in antiquity, these being the 'Bowl', the 'Pot' and the 'Shaft':

• The **'Bowl'** furnace – the earliest, most basic type. A clay-lined hollow in the ground would be filled with iron ore and charcoal. The temperature would be increased using draught from a bellows, and the furnace possibly covered by turfs to minimise heat loss. Temperatures achieved by this type of furnace are not considered to be high enough to make molten iron, but would produce a small, spongy mass of iron and slag, needing subsequent hand separation after cooling, reheating and hammering into a bloom (Fig. 87; Bestwick & Cleland, 1974, 145).

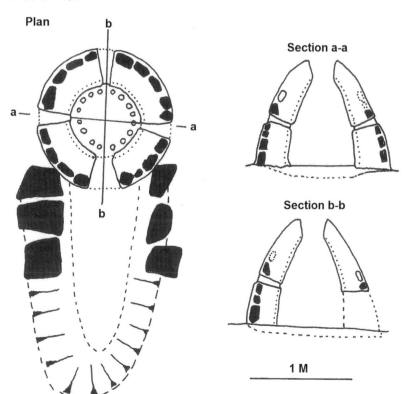

Fig 87. Developed bowl bloomery furnace from Minepit wood. (Artwork by J. R. Travis, from Tylecote, 1986, fig. 103)

- The '**Pot**' or domed furnace consisted of a flat hearth, covered by a domed framework of multiple stakes, packed with clay and stones, of 1 m average diameter. It is a type believed to have been introduced by the Romans, with examples known from Wilderspool, Pentre and The Weald dating from the first and second centuries AD. The furnace would be charged with alternate layers of ore and fuel. A hole at the top of the furnace allowed the escape of gases, and served to add more materials during the process. Draught to increase oxygen flow and raise the temperature was either natural or provided by bellows, through several inlets, or 'tuyeres' above ground. Slag produced would run off the flat hearth and out through the taphole at the base, with the bloom removed through the top at the end of the process (Bestwick & Cleland, 1974, 145).
- The '**Shaft**' furnace consists of a cylindrical shaft, built of stone, 1 to 1.5 m high, with an internal diameter of 30 cm, widening to the base. Tuyeres above ground provide draught to raise the temperature, and an opening at the base acts as a taphole to drain slag and to remove the finished bloom. Examples have been found at Ashwicken in Norfolk, Pickworth in Lincolnshire, Holbeanwood and Broadfield, both in Sussex (Tylecote, 1986, 157–158).

The furnace would be preheated for half a day, and then fuel added through the top until temperatures of 1200° to 1300° C were achieved. Then small charges of 1 kg of ore and fuel together were added to produce the bloom. Experiments by Cleere on a reconstruction shaft furnace used a mixture of iron:charcoal at 1:1 or 2:1 (Bestwick & Cleland, 1974, 143–144). During combustion the ore then moves down the shaft through three zones (reducing, solution and combustion zones). In the 'reducing zone', fuel and oxygen from the induced draught (from bellows and tuyere) reduces the ore to metal.

$$Fe_2O_3 + 3CO \rightarrow 2Fe + 3CO_2$$

This reaction releases carbon dioxide (CO_2), and slag from gangue is collected and tapped off at the base (Healey, 1978, 189, Fig. 28). A spongy mass of trapped slag and iron (the 'bloom') then remains in the shaft, the process producing up to 10–20 kg iron per day (Fig. 88; Bestwick & Cleland, 1974, 145).

Cleere, however, proposed an alternative classification to that by Coghlan, based on two main types, with further subdivisions (Bestwick & Cleland, 1974, 145):

Group A – hollow hearthed furnaces where tap cannot escape (equating to Coghlan's 'Bowl' and most 'Pot' furnaces), falling into two subtypes:
i) bowl furnaces
ii) furnaces with superstructure

Group B – flat hearthed furnaces where tap can be removed at the base (variations of Coghlan's 'Shaft' furnaces and some 'Pot' furnaces), falling into four subtypes:
1.i) cylindrical furnaces with forced draught
1.ii) conical superstructure with forced draught
2.i) as type 1.i) cylindrical but with natural draught
2 ii) as type 1.ii) conical but with natural draught

In either case, the furnace walls would have been clay lined between 10 mm and 30 mm thick. During the smelting process the clay changed into a glassy liquid, which entered and reacted with the iron ore and fuel charge, contributing to the liquid slag that runs off at the base, trapping unwanted impurities (Sim & Ridge, 2002, 51). The significant improvement to the iron production process during the Roman period, with the use of shaft furnaces, was the ability to run off this slag by-product at the base and to remove the resulting bloom without destruction of the furnace, which could then be relined with new clay, ready for subsequent smelting activity.

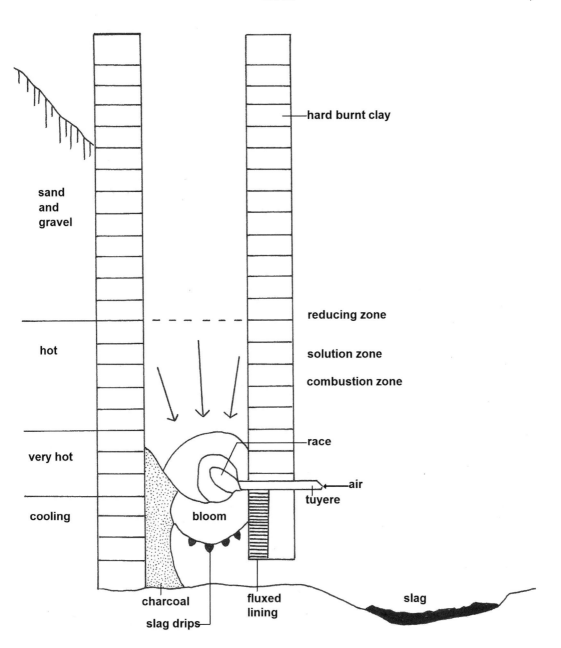

Fig 88. Cross section through a shaft furnace indicating processing 'zones'. (Artwork by J. R. Travis, from Healy, 1978, fig. 28)

The smelting process is highly fuel intensive, as it requires the production of extreme levels of heat for prolonged periods, requiring a substantial quantity of fuel. Literary evidence suggests that a variety of fossil fuels were used (peat, wood, lignite and coal), particularly from Pliny, where he describes the best wood to use for charcoal. However, translations can be ambiguous, with some confusion on his use of the word '*carbo*', which he appears to use to refer to both charcoal and coal (Pliny, *Nat. Hist.*, 16.77).

In normal usage of a shaft furnace, every 100 kg of ore produces approximately 16 kg of iron, which suggests that each furnace in operation would have needed up to 24 tonnes of charcoal per annum, requiring 96 tonnes of timber, from 3 to 5 hectares of woodland. It has been estimated that Roman Britain may have produced 2,250 tonnes of iron each year (requiring 13,500 tonnes of ore), and that smelting this using charcoal alone would have sacrificed over 240 hectares of woodland (Travis, 2005). This would have been an unfeasibly and unacceptably high level of timber exploitation and charcoal production (evidence of which would have been apparent in the landscape of the Roman period). It is therefore possible that use was made of alternate fossil fuels such as coal, selecting for use only those grades which are low in sulphur and phosphorus (which would have been harmful to the metal, causing it to become embrittled), and which can be used for both the smelting and smithing processes.

Smithing

The 'bloom' produced in the smelting furnace requires further processing, by reheating to around 900° C and hammering on an anvil, to remove the worst of the trapped excess slag, the structure of the iron itself becoming more fibrous, with layers of small amounts of residual slag inclusions ('stringers') remaining within the body of the metal. If not removed, these slag inclusions would make the metal too brittle to be of use. The quantity of the remaining slag 'stringers' and other impurities at the end of this process, visible during metallographic analysis of polished sections of samples taken from artefacts, attests the skill of ancient metalworkers and the level of purity of metal produced. These temperatures would again be achievable using charcoal, coal (if low in sulphur and phosphorus), or a mixture of both, the choice of fuel affecting the purity of the metal and the quantity of residual slag 'stringers'.

The process produces evidence in the form of fayalite slag ($2FeOSiO_2$), similar to smelting cinders but in smaller quantities. This may then further oxidise over time into soluble silica salts and iron oxides (Fe_3O_4, Fe_2O_3, or $Fe(OH)_2$) which may leave little visible remains other than slight magnetic traces (Bestwick & Cleland, 1974, 145).

This primary processing through smithing would probably be carried out close to the initial smelting site. The bloom would then be hammered into useable lumps of metal, or 'pigs', which could then be transported elsewhere if required for secondary processing ('forging') into useable objects (tools, weapons, armour, etc.).

Forging

Forging is the name given to the secondary processing of iron from the 'pig' state. This may not necessarily have been carried out at the same location as the primary smelting site, the 'pigs' of iron being of a size and form that was easily transported. This secondary processing may therefore have been carried out at civilian or military sites – either in small-scale operations, or in large, organised factories (*fabricae*). Several large-scale industrial complexes have been found in Britain, some of which were in known military locations and others that appear to have been civilian, but which may have been subject to some level of military control (for example sites at Wilderspool, Manchester and Templeborough; May 1899; Jones & Grealey, 1974; May 1922), but this is not necessarily the case in other regions of the Empire.

In the absence of buildings (workshops being often ephemeral in nature, built with the intention of being short-use, temporary and low-status structures), the main diagnostic evidence of forging may consist of hammer scale, slag (scales of iron oxide), charcoal, burnt coal, etc., possibly combined with the remains of hearths (although despite a large number of hearths having been found, not all may be contemporary, and may not therefore be representative of manufacturing methods at any specific given time).

A range of different types of hearth have been observed, however, consisting of:

1) Simple – these are small-scale, bowl-type hearths, many associated with a series of postholes, possibly serving as placements for bellows and anvils.
2) More complex oval bowl-type and L-shaped hearths – these are usually found in larger industrial complexes, possibly being used for tools and weapons.

3) 'Cell hearth' – these are also usually only found in larger industrial complexes, consisting of a square/rectangular pit, with stone benches on three sides, a sandstone block providing a base for an anvil, and a sandstone 'water bosh' (or quenching tank).

4) Pot-shaped hearth – these possibly served as a carburising furnace (with examples found at both Wilderspool and Manchester). They would have been used to strengthen cutting edges of tools and weapons through tempering with carbon (a technological advance of the late second to early third century). The initial charge would probably have been made of charcoal (Pliny refers to '*igniaria*' from mulberry, laurel and ivy), and from associated finds, the majority do appear to have been fuelled with charcoal (Pliny, *Nat. Hist.*, 16.77). However, it is possible that they may have been fuelled with coal or a combination of both coal and charcoal, although any coal used for this purpose would have needed to be low in pyrites (low sparking and low in sulphur, which would otherwise cause the metal to become brittle; J. R. Travis, 2005).

Bronze Working

Copper in its pure state is a soft metal. Although this softness may be beneficial in some cases, such as the manufacture of rivets (where it is intended for these to be periodically removed and replaced during regular maintenance), or where the aesthetic choice of colour supersedes practicalities of strength (as with early Bronze Age daggers serving ceremonial purposes), as with iron it is more often hardened by alloying with other metals, such as tin and zinc, forming the harder metals of bronze and brass (Champion, *et al.*, 1984, 215; Hodges, 1976, 64).

The processes involved in the manufacture of copper are, initially at least, similar to those of iron: mining; preparing and cleaning the ore; smelting; refining and casting; hot and cold working. It can occur naturally, as nuggets of soft native copper, just requiring minimal cold hammering, or more often is found as a range of ores in combination with other elements (most commonly as oxides, carbonates, sulphates, chlorides and silicates, and less commonly as sulphides, which are more difficult to use), the ores being easily identified by their distinctive colours and the simplest to use being the surface-weathered oxides (Hodges, 1976, 65; Table 10).

Smelting

After initial preparation of the ore (pounding and washing to remove unwanted waste products, or 'gangue', followed by sun or kiln/oven drying), the ore has to be smelted into its metallic form ('blister' copper) by heating in a furnace ('bowl', 'domed' or 'shaft', similar to those used for iron production) to around $1100°$ C in a reducing atmosphere produced by a carbon-based fuel (usually charcoal, although possibly using a mixture with sulphur-free coal), any siliceous material and other unwanted residues forming a glassy slag which either collects in the base of the furnace or is permitted to run off. The 'blister' copper so produced can then be formed into ingots (possibly for transportation to sites of secondary processing), prior to forming into sheet (useful for food vessels or armour: full products or fittings – cuirasses, shields, helmets, forming into hinges, buckles, etc.) or casting (in simple or multi-pieced moulds, or by single-use '*cire perdu*' – lost wax methods).

Smelting oxide ores:

$2CuO + C \rightarrow 2Cu + CO_2$

Basic copper oxide smelting to 'blister' copper

Where the simple oxide ores can be smelted directly after the basic preparation of pounding and washing, sulphide ores (Cu_2S) require roasting first in an oxygen rich atmosphere, to convert them into the oxide form (Cu_2O). In the case of the iron/copper sulphides, this roasting process first converts the ore into a mixture of iron oxide (FeO) and copper sulphide (Cu_2S). The copper sulphide then breaks down into its oxide form, as described above, and the iron oxide combines with the siliceous material to form fayalite slag (FeO.SiO$_2$), as formed in the iron production process (Hodges, 1976, 66).

Roasting sulphide ore:
1a) $2CuFeS_2 + 4O_2 \rightarrow Cu_2S + 2FeO + 3SO_2$
Any iron/copper sulphide breaks down to iron oxide and copper sulphide.

1b) $2\,Cu_2S + 3O_2 \rightarrow 2\,Cu_2O + 2SO_2$
Copper sulphate converts to oxide

Smelting sulphide ore:
2a) $2\,Cu_2O + Cu_2S \rightarrow 6\,Cu + SO_2$
Oxide combines with more sulphate to produce copper

2b) $FeO + SiO_2 \rightarrow FeO\text{-}SiO_2$
Iron oxide combines with siliceous material to form fayalite slag

The Romans deliberately alloyed copper with other metals, not only to produce a harder metal, but for the aesthetic properties of its appearance, the golden-yellow colour of the brass produced by alloying copper with zinc (referred to by the Romans as '*oricalchum*') being much favoured for currency and applied decoration (Pliny, *Nat. Hist.*, 34.4). Although producing a softer metal than true bronze (the latter being an alloy of copper and tin), it was easier to work, improving productivity, which may have been a factor in its introduction following the Marian and Augustan reforms of the later Republican period, although there may have been other factors at play.

As zinc was not at that time produced in metallic form (requiring distillation processing), it would have had to be added in the form of natural 'earths' such as calamine (known to the Romans as '*cadmeia*'), or as an oxide by-product from other industrial processes, Theophrastus referring to its use as a 'peculiar earth' that could 'enhance the beauty of copper', further confirmed in the writings of Aristotle and Pliny (Aristotle, *On Marvellous Things Heard*, 62, 835a; Theophrastus, *On Stones*, 49; Pliny, *Nat. Hist.*, 34.4).

Although it requires a lower level of technology, the initial production of copper (and copper alloys) is not dissimilar from that of iron. However, its subsequent secondary processing differs. As copper cools, it forms a dendritic crystalline structure (forming branching patterns, similar to those in snowflakes). During any hammer shaping, the previously randomised crystalline structure will begin to align, hardening the metal and forming weak points where the metal may split along these alignments. To prevent splitting or breakage the metal requires annealing or softening. The metal is heated to high temperatures, to reform the dendrites and randomise the crystalline structure, which will then be reset on cooling. During lengthy hammer shaping, this process will therefore need to be repeated frequently as the metal hardens to prevent spoilage.

Evidence of copper or bronze working may take the form of smelting furnaces, as described above, although it is necessary to be mindful of the similarities to those used in iron production, even in the possible presence of fayalite slag produced, or the presence of coal or charcoal fuel, common to both processes. Secondary processing may be more easily identified, with remains of casting hearths, crucibles, casting splashes, trimmed-off casting sprues (from filling funnels or 'breathing vents' in the mould), or flash (surplus metal from around joins in the mould). However, as the level of technology required for copper/bronze casting is relatively low, not all activity may necessarily have taken place within distinct and bespoke workshop *fabricae*, and casting could be carried out almost anywhere that a small, temporary hearth could be fashioned in a hollow in the ground. Although this would of course leave an archaeological trace, if this location were not within a known civilian or military site, its discovery would be purely down to chance.

In addition to the more traditional casting hearth, it is possible that Roman craftsmen

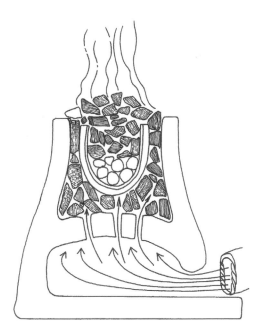

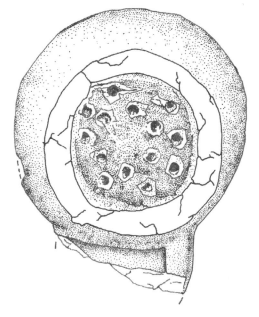

Fig 89. Schematic of the Vindonissa pot in operation. (Artwork by J. R. Travis, from Trachsel, 1997)

Fig 90. The Vindonissa pot as found. (Artwork by J. R. Travis, from Trachsel, 1997)

made use of small, portable casting hearths/pots, which could equally have been located within any workshop, or used almost anywhere for small-scale production or repairs (therefore leaving no trace of any hearth remains), as evidenced by the 'casting pot' discovered at Vindonissa (Figs 89–91; Trachsel, 1997, 141–155).

The casting hearth would probably have been fuelled using charcoal (as also would the Vindonissa 'casting pot'), coal being unlikely to have been used due to possible contamination from any sulphur present. As copper melts quite readily, at temperatures around 1000° C, this can be easily achieved using charcoal (which gives an intense but relatively short-lived burn), the combustion aided by use of bellows. Further, the charcoal being a 'clean' fuel, with few chemical impurities, it is possible to gauge the readiness of the copper mix from the

Fig 91. Reconstruction of the Vindonissa casting pot. (Photograph by J. R. Travis)

colour of the flame, which changes to a green colour when molten, at which point the mix may be augmented by the addition of other minerals to improve its properties.

If sheet metal has been used, it may be possible to find offcuts, part-formed objects or wasters, or, if objects have been lathe-spun, there may be distinct areas of copper alloy dust or small pieces of metal trimmed from the objects being shaped, indicating the location of lathes, as at Exeter (Bidwell, 1997, 90).

With the melting down of old copper alloy articles for recycling, the definition blurs between copper-tin (bronze) and copper-zinc (brass, or '*oricalchum*'), with some artefacts including a mixture of both. While it is possible that the resultant mixture in some objects is purely accidental, there is a possibility that the Roman craftsman was able to gauge the properties of the metal he was working with, to produce a mixture which best suited the artefact he intended to produce.

This view is supported by the results of examinations carried out by Bishop at the University of Sheffield, on copper alloy artefacts from both Buxton Museum and from Longthorpe, near Peterborough, to analyse their composition using atomic absorption spectroscopy (Bishop, 1989a, 11–13; 1989b, 20–22). Among the samples analysed there were examples of objects which had been cast and others which had been produced from flat plate, and although most appeared to be formed of the traditional Roman '*oricalchum*' (mainly copper and zinc), the levels of other inclusions varied, the cast objects also containing amounts of tin and lead.

Tools

Tools used during the Roman period were much the same as had been used in the fifth and fourth centuries BC, and indeed had been broadly similar from then until modern times, their function still remaining unchanged and so, correspondingly, their shape and appearance. The metalworker then and now would make use of the same basic toolset: a range of hammers (small or heavy, flat headed or round headed, similar to a modern cross-pein), tongs (to hold the hot iron), punches and drifts (to make holes in the metal), fullers (to make a piece of metal thinner), flatters (to even out the thickness after use of the fuller), swages (to mould a flat-sectioned bar into a round section), sets (a heavy 'chisel' used to cut hot bar or sheet metal), anvil and bellows (Fig. 92).

Metallurgical Analysis

Iron as a metal (ferrite) is never usually found in pure form as it almost always has small quantities of impurities present, such as phosphorus, silicon, manganese, oxygen or nitrogen. Iron (wrought iron) is more correctly called 'low-carbon steel' if

Fig 92. A selection of tools from Newstead. (From Curle, 1911, plate LXIII)

containing less than 0.5 per cent carbon, 'medium-carbon' or 'high-carbon' steel if containing 0.5 per cent to 1.5 per cent carbon, and cast iron can contain up to 5 per cent carbon (cast iron or crucible steel can be seen in use in India from the third century BC, and its origins are traditionally considered to lie in Asia, with the Romans acquiring knowledge of its production through trade contacts with India and China by around the fifth to sixth centuries AD). Wrought iron cannot be hardened by heat treatment, but can be converted to steel by 'carburisation', heating and maintaining at a temperature of over 900° C for several hours, while keeping the metal covered by carbon (coal or charcoal fuel), forcing it to slowly absorb further carbon into its structure (see Table 13).

Steel, however, is generally a deliberate alloy of iron with a greater percentage of carbon and other metals which can be hardened and tempered, for example modern 'mild steel' (only 0.25 per cent carbon, but also contains manganese) and 'stainless steel' (alloyed with nickel and chromium, and a small percentage of molybdenum). When heated and rapidly cooled, these deliberately added impurities increase the strength and hardness of the metal, but reduce the flexibility. Some 'mild' or 'natural' steels in antiquity were produced as a result of working ores containing manganese. Some deliberate steels were also produced from the early Iron Age, but these usually contained less than 1.5 per cent carbon. However, one example of a steel punch has been found from a fourth-century AD Roman context, at Heeten in the Netherlands, with an extremely high carbon content of 2 per cent, which the excavators feel may cause reconsideration of the previous view of a uniquely Near Eastern or Asian origin for this level of technology (Godfrey and van Nie, 2004). Most iron and steel was deliberately produced by the 'direct' or 'bloomery process', with the 'indirect' or 'blast furnace' processes, used for cast iron and crucible steels, generally being considered to be modern methods.

Some substances can exist in more than one crystalline form (allotropy) due to atomic changes at a defined transformation temperature, for example carbon, which can exist as both graphite and diamond. Similarly iron can exist in several forms, defined as alpha, beta, gamma and delta. At 769° C (the A2 change), the alpha iron (ferrite) loses its magnetism. At 937° C (the A3 change) the atomic structure sees a marked contraction, re-crystallisation, followed by grain growth, with a reversal at 1400° C. These changes produce heat, causing anomalies in the heating/cooling curves, known as critical points or 'A' points, occurring at slightly higher temperatures on heating than on cooling, an effect known as thermal hysteresis (Bain, 1945, 6–17).

If a molten alloy of iron and carbon (less than 0.85 per cent carbon) is allowed to cool slowly, it solidifies at high temperatures (above the eutectic temperature of 723° C, or 1000 K) to an intermediate, metallic, non-magnetic, solid solution of carbon in iron, known as austentite (named after Sir William Chandler Roberts-Austen, 1843–1902), or gamma-phase iron. It has an open crystalline structure, offering the opportunity to trap a high proportion of carbon in solution. As it cools, this dissolved carbon is ejected from the spaces in the iron lattice, forming grains of pure iron, or ferrite (Brinell hardness 80). The ejected carbon then combines with some of the iron to form iron carbide (Fe_3C), known as cementite (a brilliant-white compound with an orthorhombic crystal structure, which is brittle, but with a Brinell hardness of >600), changing the Austentite to a mixture of ferrite and cementite (Bain, 1945, 17, 74). As the mixture slowly cools to below 723° C, the iron carbide forms into alternating eutectoid laminations within the ferrite, creating at 695° C (the A1 change) a structure known as pearlite (Brinell hardness 170 to 319, depending on coarseness), taking its name from mother of pearl, which it resembles at microscopic level (Hodges, 1976, 216; Bain, 1945, 18, 23–31; see Table 14). The term 'eutectic' comes from the Greek *'eutektos'* ('easily melted') and describes a mixture of two or more solid phases which simultaneously crystallise from the molten state into a solid with a lamellar structure at a temperature known as the eutectic temperature.

At a solution of exactly 0.85 per cent carbon, however, ferrite grains will not form, with the entire material becoming pearlite. Between 0.85 per cent and 1.8 per cent carbon, grains of iron carbide form as infilling to the pearlite eutectoid, up to a maximum solubility of 2.03 per cent carbon at 1147° C (1420 K). As the mixture cools, grains of ferrite precipitate out of solution (forming ferrite layers), which as a result increases the proportion of carbon in the remaining mixture, encouraging formation of cementite, which then precipitates out, thus creating the lamellar structure as the cooling process continues (Bain, 1945, 23–31). If the alloy is artificially cooled from above the eutectic temperature of 727° C, by quenching in oil this will cause finer lamellae of ferrite and cementite to form (as the time interval will be shorter between the stages for precipitation described above), producing a material known as troostite (Brinell hardness 377), which has a more finely laminated structure than pearlite (Bain, 1945, 33, Fig. 24).

If, however, the alloy is cooled more quickly, through rapid quenching in water, the carbon may not be permitted to fall out of solution, becoming trapped inside a lattice structure of lens-shaped crystal grains, transforming into a supersaturated solid solution known as martensite (named after the German metallurgist Adolf Martens, 1850–1914), with a microscopic appearance of fine, needle-like crystals (Hodges, 1976, 218; Bain, 1945, 34, Fig. 25).

Whereas the austenite had a cubic structure, that of martensite is tetragonal, deformed by the presence of the interstitial carbon atoms, creating tension within the crystal. This deformation then inhibits the ability of the atoms to slide past each other, increasing the material hardness, an effect that may be deliberately sought on the outer edges of some objects, and is often produced during 'case hardening'. This deformation of the structure of martensite also causes it to be lower in density than austenite, producing an increase in volume on transformation, an effect used beneficially in the production of the Japanese katana sword (in the production of the katana, the blade is coated with clay on the rear edge to inhibit martensite production, the expansion in the resulting differential cooling causing the martensite in the leading, cutting edge to produce the easily recognisable gently curved shape).

In addition to being harder (Brinell hardness 400), Martensite is also brittle, and quenching often produces an overabundance, which can be adjusted by further tempering through more gentle heating (below the eutectic temperature), transforming the martensite into a less brittle material (if heated to temperatures above 450° C this would have produced a material known as sorbite, or bainite, a fine dispersion of cementite in ferrite; Hodges, 1976, 218; Bain, 1945, 44, 228, Fig. 31). This tempering process would have been controlled visually by watching the changing colour of the heated metal ('Colours of Temper'), depending on what purpose was intended for the object being produced (Hodges, 1976, 84; Table 11).

With these metallurgical characteristics of iron in mind, a metallographic examination was carried out in 2002 by Fulford, Sim and Doig, of polished sections from a range of Roman armour fragments, dating to the first to third centuries AD, from locations across Europe including Britain (from *lorica*, scale armour, mail and shield bosses), with the aim to determine, if possible, the composition, purity and hardness of metal used, the extent of use of steel, and to attempt to suggest methods of production, by measuring the thickness and mean hardness of the sample, and microscopically examining it to view the grain structure and estimated level of slag inclusions (Fulford, *et al.*, 2004, 197–220).

It was found, on samples that had not mineralised too completely to permit examination, that the majority (70 per cent of the forty-three pieces sampled) were less than 1 mm thick, with almost all (with a few notable exceptions) measuring less than 2 mm in thickness. Most of the mail samples were less than 1.1 mm diameter, with only two samples exceeding 1.2 mm (Fulford, *et al.*, 2004, 201).

It was also found that 80 per cent of the samples tested were formed from composite sheets of metal, with multiple layers (usually two or three sheets, but occasionally up to four layers), some of which (20 per cent, or seven samples in total) possibly being folded single sheets, but many (including all helmet and shield bosses) being formed from sheets of different metals, taking deliberate advantage of their different tensile properties. On examination of the grain structure in these samples, the interfaces between the grains on many were straight, suggesting that the metal had been thinned with an overall even pressure, an effect which Fulford, *et al.*, propose could have been produced by the use of hot rolling, or large-area trip hammers, the use of which (water powered) has been suggested at a late Roman mill at Ickham, Kent (Fulford, *et al.*, 2004, 201; Lewis, 1997, 111).

From the discussion of how the types of iron and steel are produced, and the appearance of the resulting grain structures within them, it was possible to identify on examination those samples that were of martensic quenched steel and others that were almost pure iron (ferrite), the grain shapes also indicating hot or cold working. Elongated grains, for example, would have formed when the metal was cold worked, the grains compressing and stretching as the metal was thinned. Where equi-axed grains were noted, this either indicated hot working or annealing, or may just represent elongated grains being viewed end on, from a transverse position on an end edge (Fulford, *et al.*, 2004, 211).

With this in mind, it was found that all helmets had elongated grain structures, suggesting cold working, without later annealing. This is what would have been anticipated considering the time and skilled labour involved in producing a helmet bowl, and the inherent difficulties in performing this task using hot metal. Annealing the metal after shaping would have softened the bowl unacceptably (with annealed iron usually being around 90 to 100 Hv in hardness). It was found, however, that quite the opposite was the case, as all helmets and shield bosses sampled had been produced from multiple layers, using harder and stronger fine-grained metal on the outer surface to withstand blows, and softer, coarse-grained metal on the inner surface to absorb impacts (helmets being harder than bosses, the former being above 210 Hv, the latter below this), indicating the superior knowledge and skill of the Roman armourer (Fulford, *et al.*, 2004, 211). In contrast, it was found that most *lorica segmentata* components consisted of mixed-grain metal, some particularly large and coarse grained, indicating a lower level of working of the metal, again as would be anticipated, flat sheet requiring less working than the more complicated shaped helmets and bowls.

Among the forty-three samples examined by Fulford, *et al.*, a number were found to have been of steel, including *lorica* and scale from Carlisle and Vindolanda, a shield boss from London, and several examples of mail from Thorsbjerg (Upper Germany) and Stuttgart (Fulford, *et al.*, 2004, 206). All of the steel samples had equi-axed grain structures, indicating heating after forging to above 723° C, to become austenite, followed by either air cooling, to ferrite and lamellar pearlite (as seen on the Carlisle scale, although as in this case only the outer surface was seen to be of pearlite, it has been suggested that this may indicate use of carburization), or water quenching to martensite (three of the five samples of mail from Stuttgart being martensite, suggesting that they had been quenched in this way, producing a hardness rating of between 125 and 275 Hv; Fulford, *et al.*, 2004, 206).

Fulford viewed the anomalies of the examples of the steel mail from Stuttgart as being possible evidence of regional variation in manufacture. This is also supported by examination of the level of slag inclusions and slag stringers in the samples, indicating the level of purity of the metal. Although remarking on the purity of all samples (with less than 4 per cent slag in over two-thirds of the samples), he noted that those from Denmark and Germany were particularly pure (at least 0.5 per cent), indicating the high level and sophistication of the technology in use (Fulford, *et al.*, 2004, 206).

This level of purity of Roman ferrous metal, and absence of high levels of slag stringers, was further remarked on by Sim in his discussion of the examination of the Carlisle armour fragments and his unsuccessful attempted duplication of similar-quality metal (Sim, 2005). However, all of his experimental samples had been produced exclusively with the use of charcoal as fuel, without consideration of alternative fossil fuels, such as low-sulphur coal, the use of which he had been unaware of in the Roman period (Sim, pers. comm., October 2005), but which would have made these levels of purity achievable. It is therefore possible that this high quality of Roman ferrous material could indirectly strengthen the argument for greater use of coal in the manufacture of iron in the Roman period (Travis, 2005).

Common name	Chemical composition	
Main types		
Cuprite	Cu_2O	Red oxide
Melaconite	CuO	Black oxide
Azurite	$Cu_3(OH)_2(CO_3)_2$	Blue basic carbonate
Malachite	$Cu_2(OH)_2CO_3$	Green basic carbonate
Chalcanthrite	$CuSO_4 5H_2O$	Blue (vitriol) sulphate
Atacamite	$Cu_2(OH)_3Cl$	Chloride
Chrysocolla	$CuSiO_3 2H_2O$	Silicate
Less frequently occurring sulphide ores		
Chalcocite	Cu_2S	Copper sulphide
Covellite	CuS	Copper sulphide
Chalcopyrites (copper pyrites)	$Cu_2Fe_2S_4$	With iron sulphide
Bornite (peacock ore)	Cu_5FeS_4	With iron sulphide
Sulphide ores naturally combined with arsenic and antimony		
Tetrahedrite	$(Cu.Fe)_{12}Sb_4S_{13}$	With iron and antimony
Bournonite	$CuPbSbS_3$	With lead and antimony
Tennantite	$(Cu.Fe)_{12}As_4S_{13}$	With iron and arsenic
Enargite	$Cu_3As.S_4$	With arsenic

Table 10. Types of copper ore found (from Hodges, 1976, 65).

Temperature	Colour of metal	Uses
220 – 250° C	Yellows	Razors, turning tools
250 – 270° C	Browns	Axes, wood chisels
270 – 290° C	Purples	Swords, knives
290 – 330° C	Blues	Saws, stone chisels, cold chisels

Table 11. 'Colours of Temper' (after Hodges, 1976).

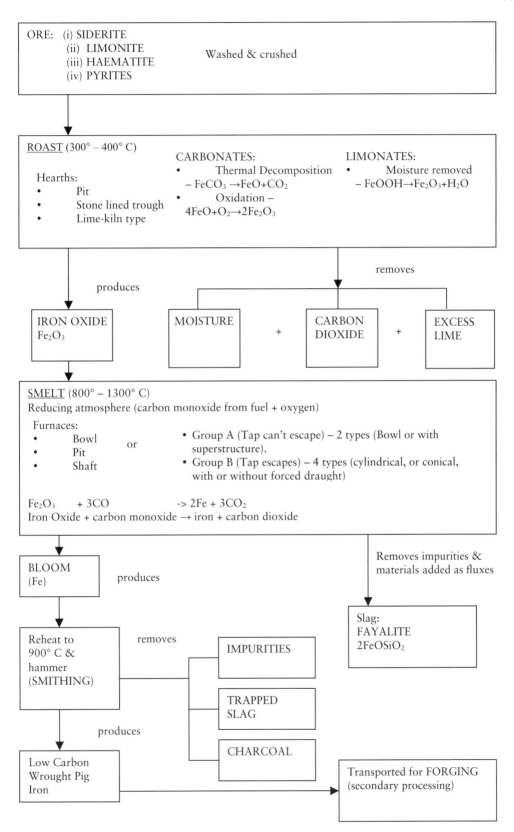

Table 12. Processing iron ore into pig iron.

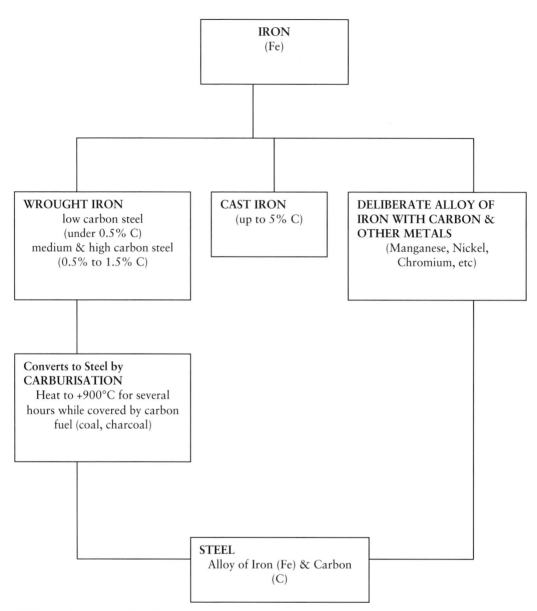

Table 13. Formation of steel.

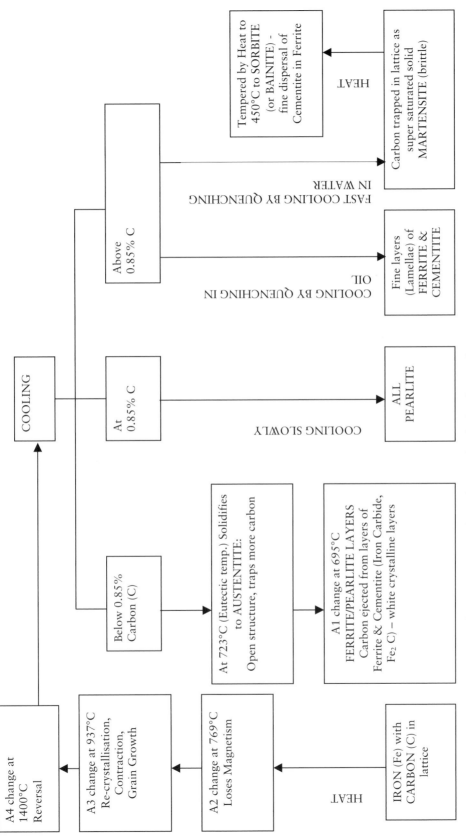

Table 14. Effects of different methods of cooling on iron, dependant on level of carbon inclusions.

APPENDIX VI

SIZE, GESSO, PAINT, PIGMENTS AND DYES

SIZE

Size is a medium traditionally applied to canvas, linen, or other surfaces intended to be painted upon, as a preparation to seal the surface of a substrate and aid in the bonding of the applied decoration. It also has the added benefit that when applied to a woven surface it causes it to shrink and tighten, providing a more even surface which can be sanded, allowing more detailed decoration. It is also one of the constituent parts in the manufacture of gesso and, in a more dilute form, provides a very strong, high-grade glue. It is ideal for woodworking applications as its solubility in hot water makes it reversible, and its generous drying time allows for repositioning (essential attributes when attempting to attach the canvas/linen cover to a shield surface). It has very little tendency to 'creep', and forms an elastic bond, which can better absorb stresses than more brittle adhesives.

CHEMICAL NAME: Hydrolysed collagen
CHEMICAL FORMULA: $C_{102}H_{151}O_{39}N_{31}$

It is essentially an animal-hide-based adhesive formed from refined animal collagen (Mayer, 1982, 259, 279–280). It is made by boiling animal skins and other organic tissue, straining to remove unwanted residues, then cooling to form crystals of pure, refined collagen. The skin-glue is then packaged and sold in this crystal form for later reconstituting and use by the artist. Before use it must then be mixed with sufficient water to achieve the correct consistency (for use either as glue or size), and heated to just below boiling point (approximately 60–63° C, 140–145° F). It should be used while still hot, as setting occurs on cooling. The bond is formed by the size/glue soaking into the wood or fabric, interlacing with the pores or threads, microscopically 'stitching' together the fibres of the two surfaces. This means that the two surfaces to be bonded must be sanded flush, and any cracks or gaps would need to be filled with some other medium (perhaps gesso for finer cracks).

Animal-hide glues/sizes can be made from a variety of animals, mainly rabbit, cow and horse, although 'rabbit-skin glue' is known to be superior for artistic purposes, having the highest gel strength, viscosity and elasticity. Bovine collagen and other hide glues are, by comparison, weaker and less flexible, but may still be used in some processes as an adhesive.

Rabbit-skin size has been widely used since the Renaissance period (and probably also since the Roman period) through to the present day (many modern artists still consider it to be superior to modern acrylic substitutes). Used as a size to prepare the canvas prior to painting, it forms a barrier to the linolenic acid found in linseed oil, which is traditionally used as a binding agent in oil paints. The glue-size itself can also be used as a binding agent when mixed with the raw pigments (as was used by Early Netherlandish artists of the fifteenth century AD, such as Quentin Matsys), producing brighter, more intense shades than were possible with an oil-based medium. However, the colours will be unstable, deteriorating over time with exposure to light. Furthermore, the collagen-based glue/size is hygroscopic in nature, absorbing moisture from the atmosphere, continually swelling and shrinking, causing the painted surface to crack. With this in mind, it is therefore all the more surprising to see the level of preservation of the Dura shields and their decoration, if this was the kind of glue used in their construction and decoration (unless, of course, they had some other superior form of glue/size of which we as yet have no knowledge). However, it is also possible that the shields were then further weatherproofed by application of some form of varnish. This may have been some form of linseed-based varnish (although with the addition of some other ingredient(s) to reduce its stickiness), as linseed, being an essential part of every horse's diet, would have been universally available across the Empire.

GESSO

Gesso, also known as 'glue gesso' or 'Italian gesso' (Gottsegen, 1993, 321), is a thick, white, paint-like medium with a consistency similar to a slightly runny soft cheese. Its name is derived from the Latin 'gypsum', which is one of its main constituents (the modern definition of gypsum being a mineral form of calcium sulphate – $CaSO_4 2H_2O$), along with chalk, and possibly some white pigment, all held in a binding agent. It is used as a primer/undercoat in the preparation of a range of substrates, including canvas and wood, prior to application of painted decoration. It acts by bonding itself to the substrate, filling any cracks (in the case of wooden panels or shields), or gaps between the fibres on stretched canvases (or canvas covered shields), providing a porous surface to which the applied paint/pigment can adhere. The absorbency of the medium derives from the chalk content (calcium carbonate – $CaCO_3$) and its flexibility derives from the properties of the binding agent used – the collagen-based rabbit-skin glue or size described above (see also Mayer, 1982, 277–288). As with the rabbit-skin glue/size, it forms a bond by soaking into the surface of the pre-sized substrate, meshing itself into the fibres of the substrate and the size coating, while at the same time filling in any gaps, cracks and imperfections between the fibres.

As discussed, rabbit-skin glue/size is more flexible than the bovine or horse-based hide glue, the latter of which may be more brittle and susceptible to cracking on flexible surfaces (like stretched canvases), but may be acceptable on the more rigid surfaces of shields (although here flexibility would be preferable, in view of the potential for impact damage). However, it is possible to increase this flexibility by forming an emulsion of gesso with linseed oil (found at all locations where horses are used, as an essential ingredient in their feed), which is known to modern artists as 'half-chalk ground' (Gottsegen, 1993, 64).

PAINTS AND PIGMENTS

Apart from the painted shields found at Dura Europos, and the small traces of paint on shield fragments from Masada, there is little artefactual evidence for painted decoration on shields. However, evidence of Roman paintings can be seen from wall paintings from sites such as Pompeii, Dura Europos, and villa sites across the Empire (such as in Spain, Italy and Britain). Siddall (2006, 18–31) discussed at length the scientific analyses of paints and pigments used at many of these locations, the most significant technique used in recent years being polarising light microscopy (PLM), which has allowed more accurate identification of materials used, based on their optical properties (Siddall, 2006, 21). The advantage of this technique is that it is a non-destructive testing method, not requiring removal of sample material, and can therefore be used on art work *in situ*, as at Pompeii.

The artwork of the Pompeii and other wall paintings above differ from that described in this work in that they are executed on plastered walls, rather than the canvas or leather-coated wooden boards of shields. It is probable, however, that similar paints and pigments would have been used, albeit possibly on a different 'undercoat' base than the gesso described above for use on shields. The Romans were also known to possess more portable artwork, in the form of Greek 'old masters' executed on wood, in similar fashion to some early Renaissance works. Although these no longer exist, due to their perishable nature, they would probably have been produced using similar methods and materials to shields, with the only notable difference being the choice of pigment, based probably on cost. For obvious reasons, when mass producing a disposable, short-life item, such as a shield, with a tendency towards a high level of impact damage, the more economical colours and shades would be the reasonable choice. The only possible occasions where the more expensive, bright colours (as listed below) would have been would be on ceremonial shields or those of high-ranking officers.

Contemporary descriptions of pigments, paints and painting techniques are provided from the work of the architect Vitruvius, from Theophrastus's *De Lapidibus* (*On Stones*) from the fourth century BC, and Pliny's *Natural History* from the first century AD. Pliny in particular discussed the range of materials used (Pliny, *Nat. Hist.*, Books 33–35), dividing these into two categories, which he described as '*austeri aut floridi*' (sombre/austere or intense).

'Florid' (Intense) Colours
Vermillion (*minium*)
Azurite (*armenium*)
Malachite (*chysocolla*)
Cinnabar (*cinnabaris*)
Indigo (*indicum*)
Tyrian purple (*purpurissum*)

'Austere' (Sombre) Colours
Ochres
Green earths
Chalks
Egyptian blue

The 'florid' or intense colours (listed above), were also the most expensive, purchased by the artist's patron at his own expense, so less likely to be found on standard legionary or auxiliary shields. The 'austere', or 'sombre' colours were the cheaper, more readily available materials (so more likely to feature on shields), including a range of ochres (providing shades from yellow, through reds, to brown), green earths, chalks, and the widely used, artificially produced Egyptian blue (Siddall, 2006, 21).

Most of these colours were produced from naturally occurring minerals coarsely ground before use, although a number were produced artificially by some level of industrial processing: some by simply heating, others by combination with other chemical additives (possibly acidic). The range of possible colours can then be further expanded by the artist mixing one colour with another. Before use the ground pigment would need to be mixed with a semi-liquid binding agent, to allow it to bond with the gesso-prepared surface (this binding agent may be 'rabbit-glue'/'size', or egg yolk 'tempura', or possibly linseed oil. Once dried, this would then require the application of some form of 'varnish' (again, possibly linseed based), in order to seal the surface and protect the paint/pigment from deterioration.

A selection of the possible colours used by Roman artists is listed below, categorised by colour (both 'Florid' and 'Austere'):

Red and Orange
Cinnabar (mercury sulphide)
Realgar (arsenic sulphide)
Red ochres (Iron oxides – Fe_2O_3) – haematite – the anhydrate of yellow ochre heated to remove the included water ($Fe_2O_3 \cdot H_2O \rightarrow Fe_2O_3 + H_2O$)
Red lead (lead tetraoxide) – result of heating white lead/lead carbonate

Blue
Azurite (copper carbonate – $Cu_3(CO_3)_2(OH)_2$

Indigo (from plants)
Egyptian Blue (calcium copper silicate – $CaCuSi_4O_{10}$ or $CaO\ CuO\ 4SiO_2$) – made by calcining copper, calcium carbonate and silica)
Ultramarine (lapis lazuli) – doubtful before sixth century

Purple
Tyrian Purple
Reddish purple – heat-treated haematite
Haematite mixed with Egyptian blue
Madder mixed with Indigo

Green
Malachite – from weathering azurite:
$$2Cu_3(CO_3)_2(OH)_2 + H_2O \rightarrow 3Cu_2(CO_3)(OH)_2 + CO_2$$
Green earth (*creta viridis*) – from two minerals: 'glauconite' and 'celadonite'
Green earth mixed with Egyptian blue
Verdigris
Copper corroded in acidic liquid (vinegar or urine)
Egyptian blue mixed with yellow ochre

Yellow
Yellow ochre (iron hydroxide – $Fe_2O_3\ H_2O$) – the mineral 'goethite'
Arsenic sulphide – the mineral 'orpiment'
Lead oxide – the mineral 'massicot'
Hydroniumjarosite ($Fe_3[SO_4]2[OH]5\ 2H_2O$)

White
China clay (kaolinite)
Montmorillonite (fuller's earth)
'Ring White' (chalk mixed with crushed glass)
White lead (lead corroded in vinegar)
Calcium carbonate (crushed limestone, chalk, mollusc shell, or bird eggs)
Magnesium carbonate (the mineral 'dolomite')
Aragonite (crushed mollusc shell)

Black
Soot
Burnt ivory and bone
Coal
Charcoal
Manganese oxide (the mineral 'pyrolusite')

PROBABLE SHIELD COLOURS

There has been much debate over recent years, speculating over the possible colour schemes that may have been used for identification purposes by different units of the Roman army, that discussion up to now being centred on clothing (what is often light-heartedly referred to as 'Tunic Wars' among re-enactment circles). Among archaeologists and historians, colour has not been considered to be of particularly high importance when compared to the interpretation of the literature, sculptural evidence and hard, physical remains (buildings and surviving artefacts). However, re-enactors and living history enthusiasts are more deeply concerned with the 'human' side of history, aiming to depict the most accurate visible interpretation possible of the persons who made use of the remaining artefacts. For them colour is a major part of that interpretation, and they understandably have often been the most vocal participants in the discussion. Without wishing to further fuel an already heated discussion, I will simply note that some have suggested the use of white/natural, uncoloured cloth (as the most economical way to consistently equip and re-equip a body of men), others have suggested the colour red (as less likely to show the wearer's blood when injured, and in keeping with safe, accepted, long-standing public viewpoint), whereas others have suggested the colour blue for those units of men serving marine functions (with the suggestion of camouflage while at sea). It has been further suggested that coloured textiles may have been used for identification on the battlefield of different bodies of men, in the possible use of different coloured neckties (*focales*).

In similar fashion (without wishing to commence a similar 'Shield Wars' debate), it is also possible that colour of shield was used as a factor in aiding battlefield identification of combat units, alongside the previously discussed legion-specific shield emblems. It is known that shields were painted for purposes of protection and waterproofing. However, shield colour could make a block of men immediately recognisable from a distance far sooner than would applied emblems, which would have required greater proximity, although emblems (and their positioning on the shield) may also have been used to indicate subdivisions within a unit.

As discussed above, a wide range of colours was available for textiles (for high-status civilian clothing, as well as for cheap, durable military tunics) and for artists (for many different purposes, including house decoration), with ceremonial shields likely to see a greater variety of shades, where expense and durability would form less of a restriction. However, for the majority of shields, which were mass-produced, short-lived pieces of equipment, it is unlikely that anything other than the most economical, readily available, earth-based or vegetable-based colours would have been used. These would be the 'austere', or 'sombre' colours listed previously, including a range of ochres (shades from yellow, through orange-reds, to brown), green earths, chalks, soot/charcoal blacks and the widely-used, artificially produced Egyptian blue.

BIBLIOGRAPHY

ANCIENT SOURCES

Ammianus Marcellinus, *Histories*, trans. J. C. Rolfe, 3 vols (Loeb Classic Library, 1935–40).

Anonymous, *Notitia Dignitatum*, ed. O. Seeck (Berlin, 1876).

Appian, *Roman History*, trans. H. White, 4 vols (Loeb Classic Library, 1912–13).

Aristotle, *On Marvellous Things Heard*, trans. W. S. Hett (Loeb Classic Library, 1936).

Arrian, *Ars tactica*, trans. P. A. Brunt (Loeb Classic Library, 1976–83).

Bible, *Genesis*, ed. J. Stirling (Oxford University Press, 1966).

Cassius Dio, *Roman History*, trans. I. Scott-Kilvert, as *The Roman History: the Reign of Augustus* (Penguin Classics, 1988).

Cicero, *Pro Rabirio Perduellionis Reo 20*, trans H. G. Hodge (Loeb Classic Library, 1927).

Claudian, *De Bello Gildonicus*, trans. M. Platnauer, as *The War against Gildonicus*, 2 vols (Loeb Classic Library, 1922).

Diodorus Siculus, trans. C. H. Oldfather, *Library of History* (Books III–VIII) (Loeb Classic Library, Vols 303 and 340, 1935).

Dionysius in Rich, J. 2007. 'Warfare and the Army in Early Rome', *A Companion to the Roman Army*, ed. P. Erdkamp, Oxford.

Herodotus, *Histories*, trans. A. D. Godley, Vol. IV (Loeb Classic Library, 1957).

Homer, *Iliad*, trans. R. Fitzgerald (Oxford University Press, 1974).

Josephus, *Bellum Iudaicum*, trans. H. St J. Thackerey, R. Marcus, A Wikgren & L. H. Feldman, as *The Jewish War* (Loeb Classic Library, 1926–65).

Julius Caesar, *Bellum Civile*, trans. J. F. Mitchell, as *The Civil War* (Penguin Classics, 1967).

Julius Caesar, *Bellum Gallium* (Gallic Wars), trans. S. A. Handford, as *The Conquest of Gaul* (Penguin Classics, 1976).

Livy, *Books I–V*, trans. A. de Selincourt, as *The Early History of Rome from its foundation* (Penguin Classics, 1969).

Livy, *Books VI–X*, trans. B. Radice, as *Rome and Italy* (Penguin Classics, 1982).

Livy, *Books XXI–XXX*, trans. A. de Selincourt, as *The War with Hannibal* (Penguin Classics, 1972).

Livy, *Books XXXI–XLV*, trans. H. Bettenson, as *Rome and the Mediterranean* (Penguin Classics, 1976).

Pausanias, *Books I–X*, trans. W. H. Jones & H. A. Ormerod, 4 vols (Loeb Classic Library, 1918–35).

Pliny the Elder, *Natural History*, trans. H. Rackham and others, 10 vols (Loeb Classic Library, 1938–1967).

Plutarch, *Camillus*, Vol. II, trans. B. Perrin (Loeb Classic Library, 1914).

Plutarch, *Coriolanus, F. Maximus, Marcellus, Cato the Elder, T. & G. Gracchus, Sertorius, Brutus, M. Anthony*, trans. I. Scott-Kilvert, as *Makers of Rome* (Penguin Classics, 1965).

Plutarch, *Marius, Sulla, Crassus, Pompey, Caesar, Cicero*, trans. R. Warner, as *Fall of the Roman Republic* (Penguin Classics, 1972).

Polybius, *Histories*, trans. I. Scott-Kilvert, as *The Rise of the Roman Empire* (Penguin Classics, 1979).

Sallust, *Bellum Iugurthinum & Catilinae Coniuratio*, trans. S. A. Handford, as *Jugurthine War, Conspiracy of Catilene* (Penguin Classics, 1970).

Strabo, *Geography, Books III–V*, Vol. II, trans. H. L. Jones as *The Geography of Strabo* (Loeb Classical Library, 1923)

Tacitus, *Annales*, trans. M. Grant, as *The Annals of Imperial Rome* (Penguin Classics, 1973).

Tacitus, *Historiae*, trans. K. Wellesley, as *The Histories* (Penguin Classics, 1972).

Tacitus, *The Agricola and the Germania*, trans. S. A. Handford (Penguin Classics, 1971).

Tarruntenus Paternus, *Digesta: Corpus Iuris Civilis*, ed. T. Mommsen, vol. 1, Berlin, 1872.

Theophrastus, *De Igne*, ed. V. Coustant (Loeb Classic Library, 1971).

Theophrastus, *De Lapidibus (On Stones)*, ed. D. E. Gichholy (Loeb Classic Library, 1965).

Theophrastus, *Historia Plantarum*, trans. A. Holt, 2 vols (Loeb Classic Library, 1916).

Theophrastus, *On Mines* (cf. *De Lapidibus*, I and *Diogenes Laertius*, V.44).

Vegetius, *Epitoma rei militaris*, trans. N. P. Milner, as *Epitome of Military Science* (Liverpool University Press, 1993).

Virgil, *Aeneid*, trans. H. R. Fairclough, Vols 63 and 64 (Loeb Classic Library, 1916).

Virgil, *Georgics*, trans. H. R. Fairclough, Vols 63 and 64 (Loeb Classic Library, 1916).

Xenophon, 'Hellenica' (*Hell.*), Books 2, 4 and 6.

Xenophon, 'Lacedemoneion Politeia' (*Lac. Pol.*), 11.5–10.

Xenophon, 'Memorabilia' (*Mem.*), 3.12.

MODERN SOURCES

Anglim, S. & P. G. Jestice, R. S. Rice, S. M. Rusch, J. Serrati, 2002. *Fighting Techniques of the Ancient World*, Kent.

Bain, E. C. 1945. *Functions of the alloying elements in steel*, American Society for Metals, Cleveland, Ohio.

Banti, L. & E. Bizzarri, 1974. *Etruscan Cities and their Culture*, Batsford.

Bergman, F. 1939. 'Archaeological Researches in Sinkiang, Especially the Lop-nor Region', Vol. 1, *Reports from the Scientific Expedition to the Northwestern Provinces of China under the Leadership of Dr Sven Hedin / Scientific Expedition to the North-Western Provinces of China*, Publication 7, Stockholm, pp. 121–124.

Bestwick, J. D. & J. H. Cleland, 1974. 'Metalworking in the North West', *Roman Manchester*, ed. G. D. B Jones, Manchester Excavation Committee, Altrincham.

Bidwell, P. 1997. *Roman Forts in Britain*, English Heritage, London.

Bidwell, P. 2001. 'A probable Roman shipwreck on the Herd Sand at South Shields', *The Arbeia Journal*, Volume 6–7 1997–98, South Shields.

Biek, L. 1978. *A first-century shield from Doncaster, Yorkshire, Appendix II: 'Examination of the Shield at the Ancient Monuments Laboratory'*, reprinted from *Britannia IX*, Society for Promotion of Roman Studies (SPRS), London.

Bishop, M. C. & J. C. N. Coulston, 2006. *Roman Military Equipment from the Punic Wars to the Fall of Rome*, 2nd edition, Oxford.

Bishop, M. C. 1985. *Proceedings of the First Roman Military Equipment Conference*, ed. M. C. Bishop, BAR International Series, Oxford.

Bishop, M. C. 1989a. 'Belt fittings in Buxton Museum', *Arma*, Vol. 1.1.

Bishop, M. C. 1989b. 'The composition of some copper alloy artefacts from Longthorpe', *Arma*, Vol. 1.2.

Brailsford, J. W. 1962. *Antiquities from Hod Hill, in the Durden Collection*, British Museum, London.

Bridgewater, N. P. 1965. 'Romano-British iron making near Ariconium', *Trans. Woolhope Natur. Field Club*, 38, II; pp. 124–135.

Broadhead, W. 2007. 'Colonization, Land Distribution, & Veteran Settlement', *A Companion to the Roman Army*, ed. P. Erdkamp, Oxford, pp. 148–163.

Bruce-Mitford, R. L. S. 1964. *Antiquities of Roman Britain*, British Museum, London.

Bruneaux J-L. & A. Rapin, 1988. *Gournay II, Boucliers et lances, depots et trophées*, Amiens.

Buckland, P. 1978. *A first-century shield from Doncaster, Yorkshire*, reprinted from *Britannia IX*, Society for Promotion of Roman Studies (SPRS), London.

Buckland, P. C. 1986. *Roman South Yorkshire: A Source Book*, Sheffield.

Burns, M. 2003. 'The Homogenisation of Military Equipment under the Roman Republic', Institute of Archaeology, University College London, in 'Romanization?', *Digressus Supplement*, 1, pp. 60–85, http://www.digressus.org.

Cagniart, P. 2007. 'The Late Republican Army', *A Companion to the Roman Army*, ed. P. Erdkamp, Oxford, pp. 80–95.

Campbell, B. 2000. *The Roman Army 31 BC – AD 33: A Source Book*, London.

Campbell, D. B. 2012. *Spartan Warrior 735–331 BC*, Osprey Warrior 163.

Cary, M. & H. Scullard, 1979. *A History of Rome*, 3rd edition, London.

Champion, T. & C. Gamble, S. Shennon, A.Whittle, 1984. *Prehistoric Europe*, London.

Cichorius, C. 1896. *Die Reliefs der Traianssäule*, Vols II (1896) and III (1900), Berlin.

Cleere, H. F. 1970. *The Romano-British industrial site at Bardown, Wadhurst*, Chichester, Sussex Archaeological Society, Occasional Paper 1.

Cleere, H. F. 1971. 'Ironmaking in a Roman furnace'. *Britannia*, 2, pp. 203–217.

Coghlan, H. H. 1977. *Notes on prehistoric and early iron in the Old World*, Oxford, Pitt Rivers Museum, Occasional Papers on Technology 8 (2).

Connolly, P. & Dodge, H., 1998. *The Ancient City: Life in Classical Athens & Rome*, Oxford.

Connolly, P. 1978. *Hannibal and the Enemies of Rome*, London.

Connolly, P. 1981. *Greece & Rome at War*, London.

Connolly, P. 2000. 'The reconstruction and use of Roman weaponry in the second century BC', *Journal of Roman Military Equipment Studies (Re-enactment as Research), Proceedings of the Twelfth International Roman Military Equipment Conference, South Shields, 1999*, 11, Armatura Press, pp. 43–46.

Connolly, P., 1977. *The Greek Armies*, London.

COTANCE, 2007. 'The Production of Leather', The European Leather Association, http://www.euroleather.com/process.htm 12/06/07.

Cottrell, L. 1992. *Hannibal: Enemy of Rome*, New York.

Coulston, J. C. N. 1985. 'Roman Archery Equipment', in M. C. Bishop (ed.), *The Production and Distribution of Roman Military Equipment. Proceedings of the Second Roman Military Equipment Seminar*, BAR International Series 275, Oxford, 1985, pp. 220–366.

Crumlin-Pedersen, O. & A. Trakadas, 2003. *Hjortspring: A Pre-Roman Iron Age Warship in Context*, Copenhagen.

Curle, J. 1911. *Newstead: 'A Roman Frontier Post and its People'*, Glasgow.

Dando-Collins, S. 2010. *Legions of Rome*, London.

Davies, O. 1935. *Roman Mines in Europe*, Oxford.

Davies, P. J. E. 1997. 'The Politics of Perpetuation: Trajan's Column and the Art of Commemoration', *American Journal of Archaeology*, 101, pp. 41–65.

De Ligt, L. 2007. 'Roman Manpower & Recruitment during the Middle Republic', *A Companion to the Roman Army*, ed. P. Erdkamp, Oxford, pp. 114–131.

De Navarro, J. M. 1972. *The Finds from the Site of La Tène, I; Scabbards and the Swords found in them*, London.

Dearne, M. J. & K. Branigan, 1993. 'The Use of Coal in Roman Britain', *Antiquaries Journal*, 75, pp. 71–105.

Demetz, S. 1998. *The Guide*, South Tyrol Museum of Archaeology.

Eichberg, M. 1987. *Scutum: die Entwicklung einer italich-etruskischen Schildform von den Anfängen bis zur Zeit Caesars*, Frankfurt, Paris, New York.

Erdkamp, P. 2007. 'War & State Formation in the Roman Republic', *A Companion to the Roman Army*, ed. P. Erdkamp, Oxford, pp. 96–113.

Erdkamp, P. 2007. *A Companion to the Roman Army*, Oxford.

Fagan, B. 2004. *The Seventy Great Inventions of the Ancient World*, London.

Fernando, D. 2007. 'Leather Trade', http://www.ferdinando.org.uk/leather.htm 12/06/07.

Feugère, M. 2002. *Weapons of the Romans*, Stroud.

Florescu, F. B. 1965. *Das Siegesdenkmal von Adamklissi: Tropaeum Traiani*, Bucharest–Bonn.

Forsythe, G. 2007. 'The Army & Centuriate Organization in Early Rome', *A Companion to the Roman Army*, ed. P. Erdkamp, Oxford, pp. 24–42.

Fox, C. 1958. *Pattern and Purpose: A Survey of Early Celtic Art in Britain*, Cardiff.

Fulford, M. & D. Sim, A. Doig, 2004. 'The production of Roman ferrous armour: a metallographic survey of material from Britain, Denmark, Germany, and its implications', *Journal of Roman Archaeology*, 17, pp. 197–220.

Gawlikowski, M. 1987. 'The Roman Frontier on the Euphrates', *Mesopotamia*, XXII, pp. 77–80.

Gilliver, C. M. 2007. 'The Augustan Reform & the Structure of the Imperial Army', *A Companion to the Roman Army*, ed. P. Erdkamp, Oxford, pp. 180–200.

Godfrey E. & M. van Nie, 2004. 'Germanic ultrahigh carbon steel punch of the Late Roman-Iron Age', *Journal of Archaeological Science*, 31 (8), pp. 1117–1125.

Goldsworthy, A. 2000. *Roman Warfare*, London.

Goldsworthy, A. 2003. *The Complete Roman Army*, London.

Gottsegen, M. D. 1993. *The Painter's Handbook; a complete reference*, New York, pp. 64, 321.

Groenman-van Waateringe, W. 1967, '*Romeins lederwerk uit Valkenburg Z. H.*', Groningen.

Gunby, J. 2000. 'Oval Shield Representations on the Black Sea Littoral', Oxford Journal of Archaeology, 19 (4), pp. 359–365.

Halpin, A. 1997. 'Military Archery in Medieval Ireland: Archaeology and History', *Military Studies in Medieval Europe – Papers of the 'Medieval Europe Brugge 1997' Conference*, 1, Dublin.

Healy, J. F. 1978. *Mining and metallurgy in the Greek and Roman world*. London.

Hodges, H. 1976. *Artefacts: an introduction to early materials and technology*, London.

Howatson, M. C. 1997. *The Oxford Companion to Classical Literature*, Oxford.

Hoyos, D. 2007. 'The Age of Overseas Expansion', *A Companion to the Roman Army*, ed. P. Erdkamp, Oxford, pp. 63–79.

Ilkjaer, J. 2002. *Illerup Ådal – Archaeology as a Magic Mirror*, Jutland.

Izquierdo, P. & J. M. Solias Aris, 2000. *Two Bronze helmets of Etruscan typology from a Roman wreck*, Nordic Underwater Archaeology.

James, S. 1988. 'The *fabricae*: state arms factories of the Later Roman Empire', *Military Equipment and the Identity of Roman Soldiers, Proceedings of 4th Roman Military Equipment Conference* (ROMEC), ed. J.C. Coulston, BAR International Series 394, Oxford, pp. 257–331.

James, S. 2004. *Excavations at Dura Europos 1928–1937, Final Report VII*, British Museum Press.

Jessop, O. 1996. 'New artefact Typology for the Study of Medieval Arrowheads', *Medieval Archaeology*, 40, pp. 192–205.

Johnson, M. A. 1930. *Etruria: Past and Present*, London.

Jones G. D. B. & S. Grealey, 1974. *Roman Manchester*, Manchester Excavation Committee, Altrincham.

Jope, E. M. 1971. 'The Witham Shield', *Prehistoric and Roman Studies commemorating the opening of the Department of Prehistoric and Romano-British Antiquities*, ed. G. de G. Sieveking, London, pp. 61–69.

Jørgensen, L & B. Torgaard, L. Gebauer Thomsen, 2003. *The Spoils of Victory – The North in the shadow of the Roman Empire*, National Museum of Denmark.

Kagan, D. & Viggiano, G. F., 2013 *Men of Bronze: Hoplite Warfare in Ancient Greece*, Princeton University Press.

Kelly, K. S. 2004. 'Imputrescible Corium: The production and structure of pre-1900 Bookbinding leather', Kilgarlin Center for Preservation of the Cultural Record, University of Texas at Austin, www.ischool.utexas.edu/~katkelly/coursework/imputrescible.html, 12/6/07.

Keppie, L. 1984. *The Making of the Roman Army from Republic to Empire*, London.

Kimmig, W. 1940. 'Ein Keltenschild aus Agypten', *Germania*, 24, pp. 106–111.

Klindt-Jensen, O. 1949. *Acta Archaeologica*, XX. pp. 1–230, Fig. 89.

Koninklijke Bibliotheek, 2007. 'Overview of leather and parchment manufacture', National Library of the Netherlands, http://www.kb.nl/cons/leather/chapter1-en.html, 18/06/07.

Krentz, P, 1985. 'The Nature of Hoplite Battle', *Classical Antiquity*, 4 (1), pp. 50.

Lancaster, L. 1999. 'Building Trajan's Column', *American Journal of Archaeology*, Vol. 103, pp. 419–439.

Le Glay, M. & J. L. Voisin, Y. Le Bohec, 2005. *A History of Rome*, 3rd edition, trans. A. Nevill, additional material by D. Cherry & D. G. Kyle, Oxford.

Lendon, J. E., 2005. *Soldiers and Ghosts: A History of Battle in Classical Antiquity*, London.

Lepper, F. & S. Frere, 1988. *Trajan's Column: A New Edition of the Cichorius Plates*, Gloucester.

Leriche P. & A. Mahmoud, 1994. 'Doura Europos, bilan des recherches recentes', *Comptes rendues de l'Academie des Inscrptions et Belles-Lettres*, p. 411.

Leriche, P. 1996. 'Dura Europos', *Encyclopaedia Iranica*, VIII-6.

Lucke, W. 1962. *Die Situla in Providence (Rhode Island): Ein Beitrag zur Situlenkunst des Osthallstattkreises*, ed. O. H. Frey, Berlin.

Luttwark, E. 1999. *The Grand Strategy of the Roman Empire: from the first century AD to the third*, London.

Mastrotto, G. 2007. 'History – Tanning', http://www.mastrotto.com/jsp/en/tanneryhistory/index.jsp, 12/6/07.

May, T. 1899. *The Roman Fortifications at Wilderspool*, Warrington.

May, T. 1922. *The Roman Forts at Templeborough near Rotherham*, Rotherham.

Mayer, R. 1982. *The Artist's Handbook of Materials & Techniques*, 4th edition, London.

Morris, P. 1979. *Agricultural Buildings in Roman Britain*. BAR British series 70, Oxford.

Osborne, H. 1970. *Oxford Companion to Art*, Oxford.

Paddock, J. 1985. 'Some changes in the manufacture and supply of Roman Bronze helmets under the Late Republic and Early Empire', *The Production & Distribution of Roman Military Equipment, Proceedings of second Roman Military Equipment Conference* (ROMEC), ed. M. C. Bishop, BAR International Series 275, Oxford.

Pearlman, M. 1966. *The Zealots of Masada. Story of a Dig*, Israel.

Piggott, S. E. 1950. 'Swords & Scabbards of the British Early Iron Age', *Proc. Prehist. Soc.*, 16, pp. 1–28.

Quesada Sanz, F. 1997. 'Montefortino-type and related helmets in the Iberian Peninsula', *Journal of Roman Military Equipment Studies*, 8, Armatura Press, pp. 151–166.

Radley, J. & M. Plant, 1969. 'A Romano-British Field System and other finds at South Anston', *Transactions of the Hunter Archaeological Society*, 9, pp. 252–264.

Rapin, A. 1983. 'Les umbos de boucliers de Gournay-sur-Aronde', *Revue archéologique de Picardie*, 1–2, pp. 174–180.

Rapin, A. 2001. 'Un bouclier celtique dans la colonie grecque de Camarina (Sicile)', *Germania*, 79, pp. 273–296.

Rawes, B. 1991. 'A prehistoric and Romano-British settlement at Vineyards Farm, Charlton Kings, Gloucestershire', *Transactions of the Bristol and Gloucestershire Archaeological Society*, 109, pp. 25–90.

Rawlings, L. 2007. 'Army & Battle during the conquest of Italy (350–264 BC)', *A Companion to the Roman Army*, ed. P. Erdkamp, Oxford, pp. 45–62.

Rees, W. G. 1995. 'The physics of medieval archery', *Physics Review* 4 (3), pp. 2–5.

Rich, J. 2007. 'Warfare and the Army in Early Rome', *A Companion to the Roman Army*, ed. P. Erdkamp, Oxford, pp. 7–23.

Richmond, I. A. 1967. 'Adamklissi', *Papers of the British School at Rome*, XXXV, pp. 34–35.

Richmond, I. A. 1982. *Trajan's Army on Trajan's Column*, London.

Rosenberg, G. 1937. *Nodiske Fortidsminder* III, pp. 1–111.

Rosenstein, N. 2007. 'Military Command, Political Power & the Republican Elite', *A Companion to the Roman Army*, ed. P. Erdkamp, Oxford, pp. 132–147.

Rossi, L. 1971. *Trajan's Column and the Dacian Wars*, trans. J. M. C. Toynbee, New York.

Russell-Robinson, H. 1975. *The Armour of Imperial Rome*, London.

Scullard, H. H. 1970. *From the Gracchi to Nero, A History of Rome from 133 BC to AD 68*, London.

Scullard, H. H. 1980. *A History of the Roman World 753 to 146 BC*, 4th edition, London.

Sekunda, N. & McBride, A., 1997. *Republican Roman Army 200–104 BC*, Osprey Men-at-Arms Series 291.

Sekunda, N. & S. Northwood, R. Hook, 1995. *Early Roman Armies*, Osprey Men-at-Arms 283.

Siddall, R. 2006. 'Pigments & Painting Techniques of Roman Artists', *Infocus*, 2, June 2006, pp. 18–31.

Sim, D. & I. Ridge, 2002. *Iron for the Eagles*, Stroud.

Snodgrass, A. M. 1967. *Arms and Armour of the Greeks*, London.

Southern, P. 1988. 'The Numeri of the Roman Imperial Army', *Britannia*, 10, pp. 81–140.

Sparkes, I. G. 1991. *Woodland Craftsmen*, Shire Album 25, Princes Risborough.

Starley, D. 2005. 'What's the point? A metallurgical insight into medieval arrowheads', *De Re Metallica: the uses of metal in the Middle Ages*, ed. R. Bork, AVISTA Studies in Medieval Technology, Science, and Art, Aldershot, pp. 207-15.

Stary, P. F. 1979. 'Keltische Waffen auf der Apennin-Halbinsel', *Germania*, 57, pp. 90–110.

Stary, P. F. 1981. 'Ursprung und Ausbreitung der eisenzeitlichen Ovalschilde mit spindelförmigem Schildbuckel', *Germania*, 59, pp. 287–306.

Stead, I. M. 1987. 'The Chertsey Shield', *Surrey Archaeol. Collect.*, 78, pp. 181–183.

Stead, I. M. 1991 'Many more Iron Age shields from Britain', *The Antiquaries Journal*, 71, pp. 1–35.

Stephenson, I. P. 2001. *Roman Infantry Equipment – The Later Empire*, Stroud.

Stiebel, G. D. & J. Magness, 2007. 'The Military Equipment from Masada', *Masada VIII: The Yigael Yadin Excavations 1963–1965 Final Reports*, Jerusalem, pp. 1–94.

Stjernquist, B. 1955. 'Simris – On Cultural Connections of Scania in the Roman Iron Age', *Acta Archaeologica Lundensia*, II.

Teixidor, J. 1987. 'Parthian officials in Lower Mesopotamia', *Mesopotamia*, XXII, pp. 187–188.

Trachsel, M. 1997. 'Ein tragbarer Giesserofen aus dem Legionslager von Vindonissa – Beschreibung, Rekonstruktion und Experiment', *Experimentelle Archaeologie Bilanz 1997*, pp. 141–155.

Travis, J. R. 2008. *Coal in Roman Britain*, Bar British Series 468.

Tuplin, C. J. 1986. 'Military Engagements in Xenophon's *Hellenica*', *Past Perspectives: Studies in Greek and Roman Historical Writing*, ed. I. S. Moxon, et al., Cambridge, pp. 37–66.

Tylecote, R. F. 1986. *The prehistory of metallurgy in the British Isles*, The Institute of Metals. London.

Van Driel-Murray, C. 1985. 'The production and supply of military leatherwork in the first and second centuries AD: A review of the archaeological evidence', *The Production & Distribution of Roman Military Equipment*, Proceedings of Second Roman Military Equipment Conference (ROMEC), ed. M. C. Bishop, BAR International Series 275, pp. 43–48.

Van Driel-Murray, C. 1988. 'A fragmentary shield cover from Caerleon', *Military Equipment and the Identity of Roman Soldiers*, Proceedings of the Fourth Roman Military Equipment Conference, ed. J. C. N. Coulston, BAR International Series 394, Oxford.

Van Driel-Murray, C. 1989. 'A circular shield cover', *Arma*, Vol. 1.2, pp. 18–19.

Van Driel-Murray, C. 1999. 'A Rectangular Shield Cover of the Coh. XV Voluntariorum C.R.', *Journal of Roman Military Equipment Studies (Spätrömische Militärausrüstung)*, Proceedings of the 11th International Roman Military Equipment Conference, Mainz, 1998, 10, Armatura Press, pp. 45–54.

Van Wees, 2004. in Rich, J. 2007. 'Warfare and the Army in Early Rome', *A Companion to the Roman Army*, ed. P. Erdkamp, Oxford.

INDEX